THE EARTH SAID REMEMBER ME

ALSO BY JASON DOVE MARK

Satellites in the High Country:
Searching for the Wild in the Age of Man

THE EARTH SAID REMEMBER ME

How to Revive Our Memories and Restore the Planet

JASON DOVE MARK

W. W. NORTON & COMPANY
Independent Publishers Since 1923

Printed in the United States of America
First Edition

For information about special discounts for bulk purchases, please contact W. W. Norton Special Sales at specialsales@wwnorton.com or 800-233-4830

Manufacturing by Lakeside Book Company
Book design by Daniel Lagin
Production manager: Julia Druskin

ISBN: 978-1-324-11788-9

W. W. Norton & Company, Inc.
500 Fifth Avenue, New York, NY 10110
www.wwnorton.com

W. W. Norton & Company Ltd.
15 Carlisle Street, London W1D 3BS

Authorized EU representative:
EAS, Mustamäe tee 50, 10621 Tallinn, Estonia

10 9 8 7 6 5 4 3 2 1

For Lily

The earth said
remember me.
The earth said
don't let go,

said it one day
when I was
accidentally
listening, I

heard it, I felt it
like temperature,
all said in a
whisper—build tomorrow

—JORIE GRAHAM, "POEM"

CONTENTS

THE EARTH SAID REMEMBER ME

INTRODUCTION

FROG IN HOT WATER

When she was growing up, Mia Monroe remembers, monarch butterflies were a natural part of the rhythm of life. This was in the 1960s, and Mia and her family lived in San Carlos, California, at that time a bedroom community of redwood bungalows and postwar cottages located south of San Francisco. She was a "nature kid," and from a very early age, she recalls, monarchs shaped her emotional landscape.

She gathered the milkweed that grew in clumps along the railroad tracks near the town's train station, then took the plants home to rear sets of caterpillars. Fall was marked by the arrival of migrating monarchs making their way to their rainy-season homes on the coast. Winter meant going with her grandmother to the famed butterfly clustering site at the Monarch Grove Sanctuary in the town of Pacific Grove.

Pacific Grove—a cute little seaside village on the southern lip of

Monterey Bay—had long been a tourist destination for monarch-spotting. A 1914 monograph, *The Butterfly Trees*, celebrated how the monarchs came to town each winter to "hang in masses from the boughs, thousands upon thousands . . . absolutely countless." There in the stands of pine and eucalyptus, Mia and her grandmother would stare upward, their eyes working to pick out the patterns of folded wings in the overstory. Whole limbs were frocked with ochre insects like coppery Christmas-tree decorations. The clusters dripped earthward: orange-and-black bangles dangling from needle and leaf.

Sometimes, as the air warmed, a group of butterflies would alight into the air all at once, a tangerine explosion. "It was amazing," Mia said. "So I grew up with monarchs. I grew up with tying them to the signals of the turning of the seasons, and to the beauty of the world, and things I did with special people."

She brought that childhood passion into adulthood. As a teenager, she began volunteering with environmental groups like the Sierra Club and then spent virtually her whole career as a park ranger at Muir Woods National Monument, the stand of old-growth redwoods just north of the Golden Gate Bridge. In the nineties, the National Park Service staff were contacted by the World Wildlife Fund and the Xerces Society, an organization dedicated to the conservation of invertebrates, to see if they could help with a census of monarchs. Mia immediately volunteered. "I raised my hand because I was already very familiar with monarchs—I was already paying attention."

Among other amateur butterfly afficionados and biologists

who were likewise paying attention, a whisper of disquiet was spreading. Monarch numbers in both the East and the West appeared to be declining. But no one was sure by how much, and scientists wanted hard numbers.

Mia and some friends and colleagues formed what would become the Western Monarch Count, a volunteer-driven, community science initiative to track the number of the monarchs that spend the winter in coastal California. Nineteen-ninety-seven marked the first year of the Western Monarch Count and it helped demonstrate that the suspicions were right: butterfly numbers were dropping. That year, volunteers counted more than 1.2 million monarchs at the overwintering sites—a figure that's never been equaled since.

During the last quarter century, Mia has watched the monarch population figures trudge downward, part of a larger disappearance in which butterfly numbers in the United States have plummeted by 20 percent. In 2000, volunteers counted some 400,000 monarchs. The following year, they counted half as many. For a while, the counts bounced between 100,000 and 200,000 butterflies. But in 2018 and again in 2019, fewer than 30,000 overwintered in California—less than 1 percent of their historic totals. At the famed monarch sanctuary in Pacific Grove, volunteers counted only 815 butterflies in 2018. The next year, the count fell by a fifth. "People like me, because I was involved in this community science effort, noticed the steady decline," Mia said. "Each of those years, I was still very worried, because I wasn't seeing them at some of the big sites."

Then the pandemic winter of 2020–21 arrived, and the butterflies did not. Across coastal California, the tallies were a string of zeroes. Zero butterflies were spotted in the Pacific Grove sanctuary. Zero butterflies were seen at Monarch Lane in the town of Los Osos, where just a few years before more than 4,000 had appeared. The youth hostel in the Marin Headlands didn't host a single butterfly. Statewide, only 1,006 monarchs were recorded that winter.

Mia was besieged by volunteers looking for solace. Her decades-long dedication and her infectious passion had earned her rock-star status in butterfly circles, and many people came to her desperate to know the reasons for the collapse. She has a weakness for jewelry (broaches, earrings) depicting monarchs and a high-pitched voice that fizzes with excitement when she talks about monarchs. Some people call her "the monarch czar."

That fearful winter, she did her best to buoy others' spirits. "I felt like I was a kind of courier sharing information," she recalled. "So-and-so planted milkweed. So-and-so is working in their neighborhood garden." Even the smallest of sightings became cause for celebration: "You begin to appreciate one or two. Each one became very special. Say—there's four! And you'd cheer on those four. You want them to represent hope for the future."

Privately, though, she was depressed. To try and stay positive, she worked in her native plant garden and did Zoom presentations for the neighborhood kids on naturalist topics. She re-read some of the classics of environmental literature: Rachel

Carson, John Muir, Gary Snyder. But having spent years, as she said, "sounding the alarm," she was now experiencing a crisis of faith. "Oh, is this what the shifting to the real dark days are? Who knew it would be this way for this beautiful insect?"

Monarchs had always been part of her life. What would the world be without them?

Then something close to miraculous happened. The monarchs came back—or at least some of them did. In the winter of 2021–22, the western monarch community experienced a surge of relief as butterflies once again swarmed to the pine and eucalyptus groves. The volunteer monarch monitors counted some quarter million butterflies at the height of the overwintering gatherings. The following winter of 2022–23 was even better: that Thanksgiving Count tallied about 335,000 monarchs in California, the highest figure in more than a decade.

Journalists who just a year or two before had reported on the collapse in the grim tones of an undertaker now rushed to write a happy story. "California's Western Monarch Butterflies Are Making a Comeback," the *New York Times* reported. "The Monarch Butterfly Beats Extinction in Triumphant California Comeback," *Los Angeles Magazine* trumpeted. The *San Francisco Chronicle* declared, "Environmentalists Cheer as Monarch Butterfly Numbers Continue to Rebound in California."

Even though Mia and other butterfly advocates were encouraged by the uptick in the monarch numbers, the public response to the tentative rebound unsettled them. The monarch population in the West was only a fraction of what it had been in the late nineties, yet the media was popping champagne corks. A decimation had been greeted with a celebration. Decades of diminishment seemed to have shrunk people's ideas of abundance and flattened their expectations of environmental wellbeing. Much of the public had lost the plot.

In conversations with her volunteers and the general public, Mia could spot people's paltry, ahistorical notions of what a robust monarch population was supposed to be.

"Yeah, there's a couple in the yard," people would say to her.

"But I got to give you difficult news," she would tell them. "It's not like it was back in the day."

And people would respond, "There's monarchs all over."

"Like, tens of thousands?" she would say.

"Oh, I saw a few."

"OK. Do you remember when you used to see ten thousand?"

There's a name for the sort of amnesia that Mia Monroe witnessed among people trying to grasp the plight of the butterflies—the way in which the present swallows the past, what it's like for destruction to blend into the background, how each of us every day grows a little more accustomed to a natural world a little less vibrant. Scientists call it *shifting baseline syndrome.*

Here's how it works. Each of us has our baseline expectations—usually forged in youth—for how the world is supposed to work, how it's supposed to look. Then the world morphs and the baseline shifts. But either we forget how the world originally appeared, or else a new generation comes along and establishes a whole new baseline based on the conditions of its youth. With diagnostic precision, shifting baseline syndrome puts a name on something that most of us have experienced—the ways in which the fog of forgetting veils experience. We live in an invisible present, blinded by the glare of now.

Shifting baseline syndrome has a clinical ring to it, like something from psychiatry's official manual of mental disorders. But it's a commonplace malady. You've experienced environmental amnesia anytime you've struggled to remember what summers were like before the nonstop, sizzling heat, in those years before fire season was an annual Armageddon. You've been caught in shifting baseline syndrome if you've ever tried to tell a younger person what a beloved place was once like—*there used to be red-winged blackbirds at this pond*—only to be greeted by a look of confusion.

Once one recognizes shifting baseline syndrome, it appears everywhere, like a pattern emerging from the forest understory. Just think of the speed of technological progress, and how difficult it is to remember social life before social media. Or maybe, to prove the point, you're young enough to *not* remember pre-social media life at all. Consider the changes in fashion, or cars, or music. Look the steady coarsening of American civic and political culture and the ways in which the once-outrageous—the bilious, hateful rhetoric and the creeping authoritarianism—have become almost ordinary.

The invisible present befuddles each generation in every age.

The passenger pigeon, the chestnut tree, and bison by the million once defined the American landscape, but they're long gone—and by now that's seemingly how it's always been. The absence just feels normal. Think about how difficult it can be to recall exactly what the farm country looked like before the sprawling suburbs shoveled it under. Baby boomers remember that they paved paradise and put up a parking lot; their Gen Z grandkids have only ever known the parking lot.

'Twas ever thus. "It never failed that during the dry years, the people forgot about the rich years, and when the wet years returned, they lost all memory of the dry years," John Steinbeck wrote in *East of Eden*. "It was always that way." The world changes and we change with it. The ancient wisdom is right: you can never step in the same river twice.

But the pathology of shifting baseline syndrome is especially acute—especially worrisome—when it comes to our twenty-first-century relationship with nature. We now live on a planet that is becoming unrecognizable, an Earth that is different from any ever experienced by humans. With that fact comes another: how easy it is, even amid global transformation, to forget the way things used to be and get used to the way things are. Change is the one constant of life on Earth; that constancy is exactly what makes change so difficult to spot.

The human aptitude for adaptability is one of our most celebrated traits. As a species we're great at adjusting to new circumstances. We accommodate ourselves to novel conditions.

We acclimate to new climates. The mind is a prediction-making organ, and as an instinctual survival technique we tailor our expectations to fit today's circumstances. We normalize and muddle through.

Adaptations, however, don't always confer advantages. When conditions change, a once useful feature can become a bug. Our knack for accommodation increasingly looks like what evolutionary biologists call a *maladaptation*. What if our habits of normalization lead us to accept super storms and skimpy winters as normal? "Man is a creature that can get accustomed to anything," the Russian novelist Fyodor Dostoyevsky once wrote, "and I think that is the best definition of him." The dour Dostoyevsky probably didn't mean this as a compliment to human nature; the line appears in a book, *The House of the Dead*, about life in a Siberian prison camp.

Shifting baseline syndrome warps our understanding of the world. It threatens to leave us—individually, collectively—unable to tell the difference between impoverishment and plentitude. The invisible present obscures how much we've lost, all that we're still in the process of losing.

The salmon: Where did they all go? The constellations of sea stars: Didn't there used to be more of them in the tidepools? We once captured fireflies in the cool summer evenings. Now, fireflies have become scarce, the July nights have grown too hot for comfort, and the summer skies are often blotted out by smoke. Or maybe it was always like this. It's hard to recall.

The concept of shifting baseline syndrome was coined in a one-page essay written in the mid-1990s by a fisheries scientist, Daniel Pauly, who was attempting to describe a collective amnesia he had perceived among fishermen. Pauly noticed that each generation of fishermen looks to the catch at the beginning of their fishing career to establish their sense of normal—without recognizing that their normal, their baseline, would be considered meager by previous generations. Pauly described this as "a gradual accommodation of the creeping disappearance of species."

For an awfully short article—the author banged it out in a single afternoon to help a colleague fill a hole in an academic journal—Pauly's essay has had a very long influence. The idea of shifting baseline syndrome has been referenced in at least two thousand research papers, on topics ranging from the perceptions of bushmeat hunters in Africa, to the quirks of cultural preservation in Australia, to the loss of ethnobotanical knowledge worldwide. Shifting baselines have been cited as a complicating factor in ecosystem management: Should we be trying to restore a landscape to how it looked a century ago, or five centuries ago?

Researchers have investigated the ways that shifting baseline syndrome influences how—or even whether—children develop a commitment to environmental protection. If a kid only knows a despoiled environment, how could they possibly be expected to realize that the pollution they're exposed to once would have been thought of as abnormal? "Across generations, the baselines shift downward for what counts as healthy nature," writes Peter Kahn, an environmental psychologist who coined the term *environmental generational amnesia* at about the same time that Pauly came up with the idea of shifting baseline syndrome.

Here's where things get more worrisome. Scholars have shown that collective landscape amnesia can occur not just across generations, but within the space of a single lifetime. One experiment, conducted in England, looked at how people understand the diminishment of ordinary bird species like the common cuckoo and the house sparrow. It concluded that some people struggle to see environmental loss because they never noticed environmental abundance in the first place. Even though bird numbers had gone down, younger study participants didn't catch onto the losses. Researchers have an unsettling term for this: "knowledge extinction."

Shifting baseline syndrome was initially considered an evocative parable of humans' estrangement from wild nature—a warning in the form of a hypothesis. At this point, environmental amnesia's implications for our ability to preserve and protect the planet have been firmly established. This isn't a drill: the syndrome is real.

"Shifting baseline syndrome is no longer a cautionary tale," one group of researchers have concluded, "but instead is a real problem for those using human perceptions of change to inform conservation policymaking." According to Kevin Gaston and Masashi Soga—a pair of academics who have researched the phenomenon more than anyone else—environmental amnesia is a universal phenomenon. It affects people in rich countries and poor countries, across age groups, throughout different cultures. Shifting baseline syndrome, they warn, is "one of the fundamental obstacles to addressing a wide range of today's global environmental issues."

Kahn, the psychologist, cautions: "It's not simply that we adapt to diminished nature, don't know it, and do fine. Rather, we adapt, we don't know it, and unknowingly do not flourish as individuals and as a species." This blindness, he has said, is "one of the central psychological problems of our lifetime." It is also among the central *ecological* problems of our time.

One of the most poignant illustrations of shifting baseline syndrome comes courtesy of researcher Loren McClenachan. She got the idea of reviewing dockside photos of recreational anglers off Key West, Florida, where for years people have made a ritual pause at the docks to have themselves photographed with their catch. Looking at photos taken between 1956 and 2007, McClenachan established that over those five decades the average length of what people considered a trophy fish shrank by more than half. The average weight decreased by nearly 90 percent. The types of fish also changed. In the 1950s, people were hooking mammoth groupers, marine goliaths that could be six feet in length. By the early 2000s, the groupers were gone, replaced by tiny snapper barely a foot long. Fish that once would've been tossed back in the sea had come to be seen as trophies. Small fry were now a big deal.

Yet although the fish shrank, the smiles of the people in the photographs stayed the same size. The ocean's abundance had been dramatically diminished, but few recreational fishermen noticed; their expectations had been downsized, too. They kept on grinning even as the sea drained of life. They had no idea what they were missing.

For a long time, these ideas about environmental amnesia and shifting baseline syndrome were mostly a preoccupation of professional scientists. They are no longer just an academic concern. As global warming and the biodiversity crisis accelerate, the concept has trickled into the public dialogue. Shifting baseline syndrome has become one of those ideas persistently tapping at the door of the zeitgeist.

In the first major magazine article about the so-called insect apocalypse, journalist Brooke Jarvis pointed to environmental amnesia as a reason why it took so long to notice the steady decline of invertebrates. "Humans are not great at remembering the past accurately," she wrote. "This is especially true when it comes to changes to the natural world. . . . The world never feels fallen, because we get accustomed to the fall." *New York Times* essayist David Wallace-Wells has warned that the effects of shifting baseline syndrome are "bewildering" and "maddening" and threaten "profound narrative confusion." Climate journalist David Roberts is blunter: "Little by little, a hotter, more chaotic, and more dangerous world is becoming normal to us, as we sleepwalk toward more tragedies."

Those of us who are alarmed by the climate crisis and the extinction emergency have often imagined (or hoped?) that there might come an environmental cataclysm that would, finally, spur civilization into dramatic action. Surely society would confront a climate or environment "Pearl Harbor moment," some unmistakable eco-crisis that would mark a clear line between *before* and *after*. The figurative asteroid, hurtling toward Earth.

But what if that Pearl Harbor moment never comes to pass? What if the alarm bells announcing the asteroid keep blaring until they fade into white noise? Perhaps pop culture's apocalyptic imaginings have misled us. We are *already* in the midst of a mass extinction event, yet the great crash in biodiversity hasn't made much of a sound–at least not one that most humans can detect.

Maybe we have grown desensitized to destruction. Maybe we'll just habituate to one catastrophe after another, until chaos takes on the dull-gray cast of the ordinary. We check the news feed and scroll from one climate disaster to the next: Paradise, Lahaina, Asheville, Altadena. After a while, the shock wears off. The world won't end with the bang of calamity, but with the whimper of forgetting.

By now many of us have had some experience with extreme weather fueled by global climate change. Hammering rain and fierce flooding, the waters washing away the long-familiar landscape. Broiling, weeks-long heat domes. An autumn bled of color, the leaves going straight to brown. But what was at first freakish becomes, over time, tediously familiar. Weather that we might once have considered strange and scary soon slips into the stream of grinding routine. The "new normal" becomes simply normal—until it winds up old hat.

Eventually, the glib quips about the new normal end up being little more than a setup for the end-times punchline. You might've seen the memes: the diners casually hanging out at cafés under a smoke-choked sky, the golfers taking their swings as the forests behind them burn. Even cataclysm can become mere scenery.

For evidence of how easy it is for a deviant normal to be redefined downward again and again, just look at the government's official definition of "normal weather." Normal, it turns out, is a moving target.

In May 2021, the National Oceanic and Atmospheric Administration (NOAA) released its updated, thirty-year temperature and precipitation averages for the United States. For the country as a whole, the new normal temperature had increased about half a degree Fahrenheit compared to the previous normal. That might not sound like that much. But compared to the original normal from 1901 to 1930, temperatures nationwide have warmed 1.7 degrees in the last century.

This rejiggering of weather recordkeeping is much more than a technical issue of meteorology. The NOAA normals are what the local television weathercaster uses to help explain if an afternoon high temp or rainstorm is to be expected—or if it's extraordinary. The definition of "normal weather" shapes how we experience the world. What would've once been considered a "heat wave" or a "dry spell" has been downgraded to average.

If normal is always shifting, it's harder to notice the changes all around us. And as "normal" becomes an anachronism, so does "extreme." We risk getting used to the extremes—the heat domes, the droughts, the brutal rainstorms—because we'll come to think of them as unexceptional. Eventually, "record-breaking" sounds like a broken record.

Federal agencies have been regularly updating the definition of weather normals for more than a century, but some climate scientists reacted to the NOAA announcement with dismay. North Carolina's state climatologist told the Associated Press, "It seems odd to still call them normal, because 1991 to 2020 was anything but normal climate-wise." Michael Mann, an eminent climatologist, said that adjusting normal every ten years "perverts the meaning of 'normal' and 'normalizes' away climate change."

Recall the smoke storms that have choked San Francisco, Chicago, Toronto, New York. The orange-red, smoke-dimmed skies felt nothing short of cataclysmic to most people. Then they faded into old news. "Are we supposed to just get used to more smoke in the sky?" Paul Bogard, an environmental writer, wondered as smoke smothered the Eastern Seaboard in summer 2023. "The answer is probably yes," he concluded as he described "another retrenchment of our experience on Earth, another instance of 'it didn't used to be that way.'" His young daughter, though, "will feel no sadness" since "she doesn't know that these skies are abnormal." If one has never known anything different, even a fire-orange sky can glow with the warmth of childhood.

Normalization of the once abnormal can happen in less than

the space of a generation. Some years ago, a group of researchers published a study in the *Proceedings of the National Academy of Sciences* which concluded that unusual temperature variations cease to spark interest on social media in as little as five years after they happen. A freak heat wave can go from *whoa* to *meh* in less than a decade. Eventually, even lethal heat will be unremarkable. We might as well start measuring shifting baseline syndrome in dog years.

I've spent nearly all my adult life, some twenty-five-plus years, engaged in the movement to protect life on Earth. As an organic farmer, I've poured blood, sweat, and tears into building the kind of world I want. As a journalist and author, I've tried to leverage language to enlist others in the efforts for ecological sustainability and environmental justice. I have marched and chanted. In my youth, I put my body against the proverbial gears of the machine and engaged in nonviolent civil disobedience. On a couple of occasions, I found myself in the back of a police car for a good cause.

And during my life, I have only (for the most part) seen the environmental devastation mount. More than half of all human-caused greenhouse gas emissions have occurred in just the last thirty-five years. The hottest year on record was 2024, beating out the runner-up . . . which was 2023. Half—*half*—of all plastics ever made have been produced just since 2000.

In my lifetime, some three billion birds have disappeared

from North America. Globally, the population numbers of wild species have collapsed by roughly 70 percent since I was born half a century ago. Between 2010 and 2019 alone, at least 160 species were wiped off the planet forever. Among them were the Pinta Island tortoise, the lesser stick-nest rat, the Bermuda hawk. Once annihilated, they went largely unmourned. Forgetting marks the final stage of extinction.

From my years of agitating for a livable planet I know that there are many drivers behind the churn of destruction: myopia, gluttony, venality, mendacity. Rank cynicism masquerades as wisdom, insisting that there is no alternative.

I've now come to the conclusion that one of the main engines of Earth's steady demolition might be something else: plain old normalization. All of it—the extinctions, the clear cuts, the wounds of the mines, the looming chaos of an atmosphere unraveled—can be traced to the ways in which we trick ourselves into believing it's not really that bad because today is only a tad worse than yesterday.

For a long time, many people believed that global civilization's laggard response to climate change and other environmental dangers was a psychological problem, a cognitive glitch. Global warming was too distant in time and space (or so the argument went) to trigger a threat response. In the era of firestorms and freak weather, however, climate change is right in our face. It turns out that a cognitive glitch *is* an explanation for inaction—it's just not the glitch we used to think.

The normalization of environmental destruction has many

roots. It is, in some ways, a natural psychological coping mechanism. Fatigued by disaster and loss, we eventually go numb. But above all, shifting baseline syndrome is a problem of memory. It's not necessarily that we cease to care, but that we slowly cease to notice—and then inevitably fail to remember. Forgetting conspires against even the strongest commitments to defend the planet.

"Our memories have been as wiped clean as the land," British writer George Monbiot says of shifting baseline syndrome. Daniel Pauly has said, "We transform the world, but we don't remember it—we adjust our baseline to the new level, and we don't recall what was there." When memory fails, perspective falters. If we mistake a grassy meadow empty of birds and bugs for a place bursting with life, it's because we lack a long enough frame of reference with which to understand what happened there. "At the end, you want to 'sustain' miserable leftovers," Pauly warns.

Amnesia accommodates us to loss. Over time, that accommodation can lead to acceptance—which is the gateway to apathy. We go from shock to shrug and then slouch to indifference. And indifference is one of the worst things that could happen at this moment in time, this hinge of history, when the fate of Earth hangs, truly, in the balance. Indifference is the gallows where collective action goes to die.

If we are to have any chance at protecting what remains of the living world, success will rest on our ability to hold amnesia at bay. To make any lasting progress in the efforts to repair, renew, revive, and rewild the planet, we'll have to relearn how to remember.

This is a book about how easy it is to forget—and how important it is to remember.

Even if you're just hearing technical terms like shifting baseline syndrome and environmental generational amnesia, I trust that you get it. None of us are strangers to what I think of as *the great getting used to.*

When I tell people about shifting baseline syndrome, they're often quick to bring up the idea of the frog in hot water. By now it's become something of an overcooked metaphor, and maybe you've heard it too. According to the story, if you put a frog into a pot of hot water, it will immediately jump out, sensing the danger. But if you put a frog into a pot of cool water on a stove and gradually turn up the temperature, the frog will stay in the water, steadily acclimating to the rising temperatures—until it's too late, and the water reaches lethal levels, and the frog boils alive.

The frog in hot water has become a pat metaphor for civilization in the climate change era, blithely cooking to death. There's just one problem: It isn't true. The frog in hot water is one of those urban myths that just won't quit.

In fact, a frog that finds itself in steadily increasing water temperatures "will eventually become more and more active in attempts to escape the heated water," as one herpetologist has explained. A curator at the National Museum of Natural History has called the fable "bullshit." In one particularly sick burn, an Australian science communicator who carefully debunked the parable wrote, "Frogs aren't economists. They're sensitive creatures."

Busting this myth leaves us with a more unsettling allegory. The frog's real-life sensitivity and its instincts for self-preservation transform the parable from a cautionary tale to a challenge. Lost as we are in the fog of forgetting, can we humans prove ourselves to be at least as smart as a frog?

I believe we can.

Since there are already enough books about the world's environmental woes, this book explores solutions. My purpose here is to focus on the antidotes to landscape amnesia. I want to understand what cures might be available to counteract the creeping acceptance to what Pauly called the "creeping disappearances." I want to understand how some people have managed to remember Earths past. This is a resistance manual for environmental defense, presented in the form of meditations and explorations.

Who has done it—where and when—and how have they done it best? What does it take to hold on to the memories of a fuller, livelier, less chaotic world? And—crucially—how can we do this not simply as individuals, but collectively, as part of a larger community of care?

To answer those questions, I've spent years shadowing people who have dedicated themselves to preserving the planet in some semblance of what they once knew. They are ordinary folks—citizen scientists, bird watchers, butterfly lovers, amateur meteorologists, fishermen. They all know that just as normalization is a predicament of forgetting, memory is a tool of ecological preservation. Memory sustains the urgency to keep the work going.

I've found that remembering often requires something on the

order of legend, so I've traced iconic tales: monarch butterflies, salmon, the bald eagle, glaciers, winter itself. I also dived deep into the field of memory studies to understand the latest findings about the workings of the mind. We swim in a sea of forgetfulness, as oblivious to environmental amnesia as a fish is to water, and modern neuroscience can provide tools to help us see more clearly.

After all the exploring, I've come to believe that a capsule prescription for landscape amnesia has four basic elements: Go outside. Bear witness. Make a record. Pass it on.

First, get outdoors. When landscape becomes the focus of our attention, something more than a backdrop, we're more likely to remember changes within it. One of the best ways to achieve such focus is by walking through the woods, or the desert, or the beach sand. Neuroscience knows that nature exposure boosts memory, which makes time outdoors an essential step to resisting shifting baseline syndrome. Embodied experiences create indelible memories, memories that can counter the normalization of destruction. The feel of earth underfoot becomes the environmental version of muscle memory.

Next, anchor those outdoor experiences in emotion. Most of what happens to us each day ultimately disappears into a void. "We actually don't remember most of our lives," neuroscientist Lisa Genova has written, pointing out that as much as 97 percent of our experiences eventually fade. What about the other 3 percent? Much of it is what's called "episodic memory," and most of what we remember about episodes is how they made us feel.

Emotions etch themselves in the mind unlike anything else. Our moments of loss and love in the natural world, of fear and joy, create anchors in the mind.

The other kind of memory is what's known as "semantic memory"—facts and figures—and here notes can help. You don't need to catalog the appearance of each bloom with the exactitude of Thoreau or note daily temperatures with the precision of the National Weather Service. A garden log or nature journal will do. Photos will too. Visual records of what the shoreline once looked like or where the woods used to stand can provide incontrovertible evidence of change over time.

Then remember to pass it on to the next generation. Talk about the weather. Tell friends and family (especially the little ones) about the changes you've witnessed. Each of us can be environmental storytellers and authors of natural history, passing down the stories of the places we inherited.

Maybe these treatments sound familiar. They are, after all, among the old-time home remedies for our environmental ills. Now, they are invested with new urgency. Especially in the epoch of the Anthropocene—this "human age" marked by the hyperspeed of Earth's disfiguration—ecological memory is a check against the bewilderment of head-spinning environmental changes. Memory is indispensable to any endeavor to sustain a livable planet and, above all, to restore it. To remember a more abundant Earth—forests and plains full of wildlife, seas filled with fish, icy Decembers and temperate Mays—is to fix a stable baseline from which to measure progress.

In a time in which the powers that be are consciously waging a war on history—when some forces seek not just to rewrite history but to wipe it out—memory is more than mere recollection. Memory is an action infused with political importance. Amid calculated forgetting, remembering is nothing less than a form of rebellion.

Ecological memory alone won't deliver us from our environmental crises. There's no going back to the past. Once the damage is done—the mountaintop blown to pieces to make way for the coal mine, the atmosphere twisted beyond recognition—it's done for a long time. But memories can help to mend the future. To be good ancestors, we'll have to remember that memory is the fuel of imagination.

Shifting baselines have set us adrift. Remembrances of Earth can be a common shore, a touchstone from which to navigate the uncharted waters of this hot, chaotic, fast-moving century.

CHAPTER 1

GO OUTSIDE

My own personal efforts to resist environmental amnesia are complicated by an inconvenient fact: I have the ecological baseline of a kindergartener. I've passed the half-century mark, but I've only lived here in my home place—the landscape to which I've hitched my heart and where I expect to spend the rest of my days—for a short time. My baseline is barely five years old. Yet this is also a blessing. It's forced me to learn—*remember*, you might say—that the task of creating memories of the living world takes work, demands presence and attention to time and place.

My oldest environmental memories are anchored in the Sonoran Desert where I was born and raised. What I remember most of all is the smell of creosote after a rain. The creosote is a gangly shrub, as awkward as a teenager, its tiny green leaves sticky with oils. When a rain strikes, those oils blast out a cloying aromatic cocktail of sandalwood meets grapefruit meets tobacco.

After the unbroken aridity, the desert turns effervescent. Imagining the smell takes me back to palo verde trees sketching the routes of the arroyos, saguaros standing sentinel over the scene.

Most of my life, though, has been spent in Northern California. It took me some twenty years to truly know the place. At first, all I understood was the metronome ticktock between wet season and dry season—the oak savannah Ireland-green in winter, then fading to golden-wheat by summer. Over time, as observation ratcheted curiosity, details came into view. I became acquainted with the common and the abundant: coyote brush and chamise, ceanothus (aka California lilac) whose blue flowers, if wetted and rubbed together, foam into a soapy lather. I got down on my knees to get intimate with the wildflowers: vestment-purple shooting stars, yellow fiddlenecks, white milkmaids ushering spring. Eventually, it was no problem to tell a live oak from a valley oak from black oak from a blue oak. I was at home.

Then, in the middle of mid-life, my family and I made a leap out of California and relocated nearly a thousand miles north to the northernmost corner of the Pacific Northwest, the heart of greater Cascadia. My partner's parents live here, and the magnet of family was pulling us away. But we also left California for some of the usual reasons: the high cost of living, the crush of traffic, the disillusionment of watching the San Francisco Bay Area's freewheeling spirit get swallowed into Silicon Valley's insatiable maw.

Something else was at work—a desire to raise our daughter in a landscape that felt just a tad more resilient to the ravages of climate change than drought-wracked California had come to feel. Some place cool, green, and watered. We picked up and packed out during the pandemic summer of 2020, and as we headed northward California burst into flames behind us. "You got out just in time," a buddy texted as a wildfire approached his ranch. "California is imploding." It would be too dramatic and self-mythologizing to call ourselves climate refugees, but climate *emigrants* wasn't far off the mark.

Once settled in Cascadia, I had to learn the land all over again.

Having spent a fair amount of time at the in-laws' house, I already knew the northwest arboreal trinity of cedar-fir-hemlock. In the understory, I could distinguish the smooth oval leaves of salal from the spiney oval leaves of Oregon grape. I could tell you what a sword fern was.

But mostly the woods were a green blur. *So. Much. Green.* The Cascadian rainforest is dark and dank, claustrophobic and closed in, a shrouded, uninviting place. Every root, branch, and rock is cocooned with moss—moss swaddling limbs, moss smothering stones, saddling stumps, knuckling burls. On the swooping boughs of the big leaf maples licorice ferns grow on moss, the fronds dressing the bark like dinosaur feathers.

The forest felt decadent, a riot of too-muchness. The land was illegible.

So I set out, one more time, to try to pick up the peculiarities of place. I hiked a lot, in all seasons. I bought a pile of field guides and marked them up. I took lots of pictures on my phone and uploaded them to iNaturalist and made birdsong recordings with Merlin. As much as figuring out the grammar of the woods required attention, it took patience. Attention, I learned, comes from *attendance*, from the Latin *attendere*, "to listen carefully," "to stretch toward."

I took the etymological advice, I listened carefully, and I took my time. I was childlike in my inquisitiveness, and the discoveries piled up like presents under the Christmas tree. The pearl-cluster blooms of osoberry, I observed, herald the end of the seemingly endless Pacific Northwest winter. Then spring comes like a meteor shower. White trillium and bright pink currant first, followed by red bleeding hearts shaped like valentine drawings. Tiny, white thimbleberry flowers lining the trails like twinkle lights. Midsummer is marked by wild huckleberries at sea level and wild blueberries in the mountains. When I see the maples taking on a white cast of rust, I can guess that fall is coming back around. Then the cottonwoods flip to yellow, and I know it's so.

In becoming intimate with place, I was following a well-trod path. Ever since Henry David Thoreau slapped together a shack a short walk from his Massachusetts hometown, a righteous parochialism has fueled citizens' efforts to preserve the living world, a desire to protect *this forest* or *this seashore* or *this river*. "Each place is itself only, and nowhere repeated," wrote the late, great Barry Lopez, a person who cultivated a sense of purpose from his sense of place, who made place-ness an end-in-itself. To practice that kind of radical here-ness inevitably leads to affection, and affection to allegiance. To know your personal patch of

Earth—whether you've spent your entire life there or, like me, are a newcomer—is to care for it, to care enough to defend it.

My desire to wring meaning from place was also pegged to something else. I wanted to form a mental baseline that could withstand the fast-churning dislocations of this unprecedented era. "To have any kind of home, you must have at least a little history," environmental journalist Madeline Ostrander writes in *At Home on an Unruly Planet.* Until I had accrued some personal history earned by attention, I would be, in a sense, homeless.

The weather wasn't helping. Even as I got to know the landscape, the atmosphere befuddled. My family's first calendar year in Cascadia was total climate whiplash, one season after another marked by some crazy superlative. That summer, we suffered the horrific, now-infamous heat dome during which a billion sea creatures (yes, you read that right) broiled to death amid triple-digit temperatures. That fall delivered the worst floods in our county in generations; water sloshed around the basement of our new house, forcing costly repairs. Winter came with some of the coldest temperatures in decades.

Our third spring here was one of the driest on record in Washington State, with stream flows just a quarter of normal, the first of several consecutive years of official drought. To my eyes—Arizona-raised, California-adapted—the mossy woods still seemed awfully moist and green. But what did I know? I just got here.

Climate chaos had handicapped my already tiny frame of reference. The weather patterns had gone squiggly. How could

I possibly form a baseline when the atmosphere was zigging and zagging?

The only chance of doing so rested on making direct contact with the landscape. To know and remember the living world requires immersion in real life. The first, the essential, the inescapable antidote to environmental amnesia is simply this: Go outside.

Funny, isn't it, how sometimes new ideas emerge all at once from separate places, as if a premonition were lurking in the ether of the zeitgeist? Around the same time that Daniel Pauly coined the term shifting baseline syndrome and Peter Kahn came up with the notion of environmental generational amnesia, another thinker hit on a turn of phrase that weaves, double helix-like, with those other diagnostic terms: *the extinction of experience.*

The alliterative idea was conceived by an esteemed lepidopterist—a scientist that studies butterflies and moths—named Robert Michael Pyle who was seeking to put words to the steady reduction in people's "close and intimate contact with a particular patch of ground." His memoir, *The Thunder Tree*, traces the losses he witnessed in the second half of the twentieth century as the outdoor playground of his baby boomer youth—the horizon-open high plains of Colorado—fell to the creeping sprawl of greater Denver. He personally noticed that once-common species (birds, butterflies) had become so rare in the spanking-new suburbs that it was like they had disappeared. "If a species becomes extinct within our own radius of reach," Pyle wrote, "it might as well be gone altogether."

Ultimately, Pyle warned, those local extirpations would distance people from nature. "I believe that one of the greatest causes of the ecological crisis is the state of personal alienation from nature in which many people live," he wrote. "People who care conserve; people who don't know don't care. What is the extinction of the condor to a child who has never known a wren?"

At this point, those words, published in 1993, seem almost quaint—another example of shifting baseline syndrome, of how every loss narrows our expectations of the present and fetters our aspirations for the future. In the thirty-plus years since, the extinctions and extirpations have grown while our experiences of the natural world have shrunk. We've gone from being an indoor species to becoming a digital one.

You may have heard of "nature deficit disorder," the widening chasm between people, especially children, and the living world. Today, the average American child spends a scant thirty minutes a day outdoors. They spend fourteen times as many minutes in front of screens—some seven hours daily. Even if those statistics are familiar, I hope they still manage to shock. A capacity for outrage is proof that one hasn't yet succumbed to normalization.

In the climate era, the extinction of experience threatens to become a negative feedback loop: the less time we spend outdoors, the more the outdoors becomes alien to us, which can fuel a vicious cycle of biophobia. A natural world made increasingly dangerous by climate chaos is an unwelcoming place. Who wants to hike through a forest that's been charred black by a wildfire? Who wants to risk Lyme disease for a stroll through the woods? Better to stay inside and turn on a nature documentary. No one is immune; according to one study, field biologists are spending less time outside than they once did.

Even summer camp—once the very pinnacle of many kids' experience of outdoors freedom—has retreated indoors. In recent years, extreme heat and unhealthy smoke levels from wildfires have forced some American summer camps to either move activities inside or cancel sessions altogether. "We just didn't want to put our kids at risk of getting sick," one summer camp administrator said. In tinder-box landscapes, the ritual of the campfire has been replaced by virtual fires made of flashlights and headlamps. What is a s'more to a child who has never known a real outdoor fire? Just a relic from another age. All generational amnesia begins with individual forgetting.

This enclosure of experience—if I can twist Pyle's original—exacerbates landscape amnesia. Shifting baseline syndrome is a process of acclimating to the abnormal, and a life lived indoors accelerates acclimation. It's hard to feel climate chaos if you spend 90 percent of your time (as is the case for the average American) insulated in a climate-controlled room. Sure, one can follow the destruction via a high-definition TV—and in the middle of a smoke storm, that's probably a smart call. But absent direct physical experience, the loss and damage will remain abstractions. An indoor species will miss what's happening to all the other species who live outdoors.

What is to be done? Pyle offers a straightforward suggestion: "Every day, come to your senses."

If that sounds elementary, that's the point. To resist environmental amnesia requires connecting to the elements—the sting of cold, the slick of sweat in the heat. Landscape amnesia is a matter of normalization, and another word for normalization is desensitization. Sensory experiences, embodied experiences, are part of the work required to preserve the body of the living world.

Academics who have studied shifting baseline syndrome are emphatic on this point. A few summers ago, I had a long conversation with University of Exeter ecologist Kevin Gaston who, along with his University of Tokyo collaborator Masashi Soga, has done the deepest research into the causes and consequences of environmental amnesia. Gaston was insistent that there's no substitute for time outdoors.

"Your own experiences are going to shape your views and your concept of what is and what's not normal," Gaston told me. "You're going to get the fastest changes in people's conception and understanding if you actually show them real nature." He went on, "I think we perhaps relied too heavily on the glossy magazine that educates you a bit and shows you what nice places look like, and too little on actually giving people nature experiences."

Without those direct experiences—and absent the effort of attention—it's all too easy to experience the environment as a green blur. Many of us suffer from what researchers call "plant blindness" or, more broadly, "nature blindness." For most of us in wealthy, industrialized countries, the living world mostly serves as a backdrop to the other dramas of our lives. Nature is mere diorama. This may even be true of hard-core outdoor adventurers. Do you know the names of the trees that speed past on the ski run, the species of the sea creatures that live beneath the surf waves? Or is it all just scenery?

The dilemma should be obvious enough. It's impossible to notice the disappearance of a plant or animal we never recognized. We can't possibly remember what we didn't know the names of in the first place.

The good news is that you can go outside no matter where you live. When it comes to learning the lexicon of landscape, the environment you live near trumps the epic wilderness you visit once a year—even if the local ecosystem resembles what Pyle called "leftover landscapes" and "hand me down habitats." The capital-*N* nature of national parks and wilderness areas might overwhelm with the sublime, but the nearby nature of the city park is more likely to, over time, forge bonds of attachment. Ecological memories can be made in all sorts of locations. The scrubby woodlot, the urban waterfront, and the meadow at the edge of the mall can each be places for creating one's own personal ecology, an individual environmental baseline. Sit in the backyard or on the front stoop or on the fire escape and take notice of the more-than-human. Even a golf course will do in a pinch. Anyplace is somewhere.

In that sense, familiarity with the common beats the amazement of the Instagram-worthy vista. Small, everyday interactions with plants and animals and weather will mortar an intimacy with nature. To create environmental memories, regular practice is best—visiting the same landscape week after week, season after season. Stroll the sand, walk the creek, run the trail, swim the lake, or just sit on a park bench and watch the pigeons make a mess of the grass. There's no doing it wrong.

My own experience has proved as much. By the standards of, say, Chicago or Phoenix or Atlanta, the 745 protected acres of shadowed, moss-clung forest behind my place is impossibly wild. But it's still just a municipal park, reliably busy with the human circus: mountain bikers, parents with strollers, joggers with their earbuds, packs of dudes zooming past on Onewheels, tunes blaring. The drone of the interstate is rarely out of earshot. And still. This mix of the wild and the urban has proven to be more than enough for me to—slowly, slowly—pick up on the natural patterns of my home place. Over time, the green tunnels of the woods have been graffitied with memories, those memories then inscribed with personal meaning, all of which eventually creates a sense of place.

Maybe this seems remedial, but don't be fooled—such practice takes real effort. If "practice" sounds too much like yoga or meditation, then just think of it as *habit*—as simple as, say, taking the dog for a walk to the same place every day. Habit and practice can pull landscape from the background to center stage. It reminds me of a quote I once read from a social justice nun, Sister Judy Vaughan: "You don't think your way into a different way of acting—you act your way into a different way of thinking."

Luckily, we have instinct and intuition going for us. When it comes to being outside, we're hardwired to remember.

Time spent outdoors in natural settings is good medicine for body and mind. In the last twenty or so years, neuroscientists and psychologists have scaffolded that folk wisdom with the confirmations of research. Scientists have shown that being in natural areas lowers heart rate, reduces stress and anxiety, increases people's subjective sense of well-being, and even encourages cancer remission. The memes are funny because they're true: nature is healing.

Being outside has also been linked to improvements in cognitive functions, including the workings of memory. One of the most widely cited studies on this comes from psychologist Marc Berman, who is a lead researcher at the University of Chicago's Environmental Neuroscience Lab. Berman and collaborators have demonstrated that time in natural landscapes—compared to urban areas—contributes to a marked improvement on memory tests. How come? The prevailing theory is that natural landscapes allow the brain to enter a state of "soft fascination" that helps restore the mind's ability to focus. In comparison, urban settings—with their racket and traffic—puts great demands on our attention. A parallel hypothesis is that the brain is reacting to the "visual fluency" of natural scenery. Natural patterns are simply easy on the eyes, which helps grease the mind. "In general," Berman and collaborators conclude, "cognitive tasks that require working memory and cognitive flexibility improve most reliably after nature exposure."

Then there's the proven link between memory and walking,

which is highly recommended for forestalling dementia. Walking in a natural setting seems to provide more of a cognitive boost than sitting in the same place. Walking's contribution to memory is due in part to the physical structure of the brain. The seat of memory is an area of the brain called the hippocampus, a seahorse-shaped zone located in the middle of the noggin. The hippocampus—one of the evolutionarily oldest parts of the brain—is also the center of spatial reckoning. It doesn't just remember old routes but also plays a crucial role in navigating pathways we've never taken before. The hippocampus is our internal compass—the mind's GPS—always orienting the body in time and space.

This overlap between memory and spatial-temporal orienteering makes for a potent combination. The mind is a cartographer of not just space but also experience. Memory works as a mapmaker, constantly constructing maps of the physical world as well as maps of ideas, social relations, episodes, feelings. The mind maps meaning onto space—maps layered upon other maps.

The ancients intuited this. One of the oldest mnemonic devices is what's called a "memory palace"—or the method of *loci*—a technique developed by a Greek poet in the fifth century BCE. Let's say you want to encode some fact into memory. To make the information stick, conjure up a mental image of the fact and then place it into a familiar place—actually envision putting it there. *Boston Tea Party of December 1773 + image of tea bag + the tea drawer in the kitchen* = a fact stored in memory.

I've experienced some of this myself. I don't feel like I have an exceptional memory, yet I can recall, with stunning vividness, just about every wilderness trail I've ever tramped in a quarter century of outdoor adventures. By this I mean I can see the spatial details of the trail down to a resolution of about a half mile or so. I can reconstruct in my mind's eye where the route climbed uphill in a switchback or where it crossed over a river. I can remember particularly interesting trees or specific waterfalls and their relation to other points on the landscape. All of this I see laid out on an imagined map in my mind. I have developed an "archive of the feet," to borrow a line from Simon Schama's *Landscape and Memory*.

What kind of magic is this? It seems that more is at work than just nature's properties of relaxation and the hippocampus's combination of navigation and memory.

Take "the Proust effect." A classic example of the mysteries of memory comes from Marcel Proust's novel *In Search of Lost Time*, when the protagonist dips a madeleine cookie into a cup of tea and the scent yanks him back in time to childhood Sunday mornings at his aunt's house in Normandy. Modern neuroscience has confirmed that scent memories are especially strong—and especially emotionally resonant—because the olfactory organs have a direct connection to the hippocampus as well as the amygdala, the area of the brain that processes emotions.

Science also knows that multisensory memories that combine a visual cue and an auditory cue are more durable than just a single-sense memory. "The more senses that can be recruited," neurologist Richard Restak reports, "the more likely you will be able to form a long-lasting memory. The reason? More areas of the brain are involved."

We say that "a memory leaves an impression" and this is literally true, as memories are physical things, patterns knit across the brain. "Engrams," they are called, the networks that constitute our memories. Memories are, above all, associative. They make and are made of linkages. Memories encode experience by forming connections and noticing relationships among events, people, places, things—other beings, too.

Patterns, networks, connections, relationships: there's a resonance between how the mind works and how the living world operates. Just like memory, Earth's remnant precincts of wild nature brim with interconnections. What is nature but a web of associations? Plants feeding off sunlight and soil, insects feeding off the plants, the birds feeding off the insects, each of those beings ceilinged by the trees that live off the dead bodies of all of the above, including their own grandparents.

Could it be that one of the reasons why nature benefits memory is that it excites all the senses? The outdoors, after all, is sensually immersive in a way that the indoors can't match. And all memory begins with perception: "Remember that without the senses there is no memory," the philosopher Voltaire wrote. Everything you remember began as a physical sensation—a sight, a sound, a smell, a taste, a touch. I can't help but recall my scent memory of creosote after a desert rain, and how it never fails to drag me back to the original baseline of my own personal ecology. Truly: the body remembers.

If you want to take a dose of nature and just chill, then by all means go for it. But if the goal is to form specific long-term memories that can short-circuit environmental amnesia, then something stronger is required. The best way to apprentice to ecological memory is the same method for apprenticing to any other craft: focus.

Attention is everything. "My experience is what I agree to attend to," the pioneering psychologist William James wrote in the late nineteenth century. As far as our minds are concerned, the phenomenon that passes unnoticed might as well not exist. But attention is one of the hardest things, especially in an age in which we carry in our pockets a gadget that can access endless entertainments and the entirety of human knowledge. Long before the arrival of the smartphone, James underscored the difficulty of attention; to focus on any one thing requires an act of will to shut out other things. "Attention . . . is the taking possession of the mind, in clear and vivid form, of one out of what seems several simultaneously possible objects or trains of thought."

If attention is the cornerstone of experience, it follows that attention is also the keyhole of memory. "How do we make memories that stand out in our cluttered minds?" neuroscientist Charan Ranganath writes in *Why We Remember*. "The answer: attention and intention." This takes work. "Paying attention requires conscious effort," Lisa Genova declares in *Remember*. "If you want to remember something, you have to turn your brain on, wake up, become consciously aware."

How can we attune our attention to the living world? While it may be true that there's no doing it wrong, there are ways of doing it better, doing it best. For starters, take the earbuds out, as part of a larger practice of immersion. Next, go slow. Walking is better than running or biking; for the most careful observation, standing still beats moving. Train the mind upon some specific thing, because attention operates best when it can anchor to a subject.

Above all, be present. Presence is the only way to transmute a moment into a memory, to take the present and cast it forward into the future, where it can form a record of now. Presence is the lens of attention.

I can guess what you might be thinking—this all sounds suspiciously like bird-watching.

There are good reasons why bird-watching is one of the most popular nature-related activities and why it makes such a perfect activity for forming ecological memories. Birds are beautiful and easy to spot. With their athletic acrobatics, birds are impressive. Birds are entertaining. They are comical, opinionated, disputatious, their relations as packed with intrigue as a Shakespeare drama.

More than anything, birds are neighborly. For most of us in the industrialized world, birds are the wild creatures we are closest to, since they inhabit a universe that we earthbound bipeds make no use of. Birds excel at turning the backdrop of nature into the foreground of attended experience. To make eye contact with a great blue heron or a great horned owl is to experience an unforgettable moment of frisson; it's like communicating with an ambassador from another civilization.

Not sure what a blue heron looks like? No worries. Forming meaningful ecological memories doesn't require being a black-belt naturalist. If anything, it's probably better if you're not. Discovery resides in mystery.

A beginner's mind—the kind of childlike curiosity I came into Cascadia with—is an ally in the act of memory-making. To be a good ecological memoirist demands inquisitiveness more than expertise. Learning something new triggers a release of dopamine, the brain's feel-good reward neurotransmitter. "Curiosity improves memory for both the mundane and the interesting, whether you're eight or eighty-eight," Ranganath reports.

Ideally this work of tuning attention to nature starts at closer to eight. On this point, the scholars of shifting baseline syndrome are just as insistent as they are on the importance of time outside. The most durable personal ecologies and most stable ecological baselines will be formed in childhood. The best way of combating the extinction of experience is to get 'em while they're young.

But it's never too late to start. An inspiring example of developing a beginner's mind late in life comes from the writer Amy Tan. As a novelist—she wrote *The Joy Luck Club*—Tan has been a lifelong noticer of things. Not until she was sixty-four, however, did she notice birds. When she did, it blew open a door to a whole realm she hardly knew existed. Discovering birds allowed her "to return to childhood wonderment, when everything was seen as new," she writes in *The Backyard Bird Chronicles*, which is beautifully illustrated with her loving pencil portraits of birds. "In these words and drawings are what changed me when I was naïve and curious and wondered about a bird I saw for the first time."

To encode the details of bird species into memory, her naturalist mentor, John Muir Laws, encouraged her to recite aloud the physical characteristics of birds upon the first few sightings. I love that: remembrance as incantation. I love, too, how Tan went no further than her backyard to find an undiscovered country; most of her sightings happened while sitting at her dining room table or while brushing her teeth. *Anyplace is somewhere*—including the bathroom sink.

Like Tan, I didn't get into birds until well into adulthood, in my case in my mid-thirties. Unlike Tan, whose dedication is near-obsessive (as she admits), I am a lackadaisical birder. I make an express trip from home to watch birds no more than two or three times per year; I can be inattentive about replenishing the birdseed and the suet. Yet even this modest attunement has helped me to mark and measure the living world.

Just like my topographical recall of the wildernesses that I have hiked, I have developed detailed memories of my first sighting of bird species. I remember, of course, my "spark bird"—the bird that first keyed me into the avian universe. They were the red-wing blackbirds that roosted in the tule reed–ringed pond at the farm I used to manage. Their name was easy to remember; they are the black birds with red feathers on their wings. Mostly what I recall is when I first noticed their singing. Their call is tattooed on my brain: a high-pitched trill-whistle that is so sharp it skates the edge of irritation. In addition to that spark bird, I also remember the exact place and season when I saw dozens of other species for the first time—and I mean *saw*, not spotted, the moment when I really recognized a species' individuality, saw it as the being-in-itself.

Clark's nutcracker: summertime, July, the Sierra Nevada's Emigrant Wilderness, near the crest of Mosquito Pass, flocks of them scouring the Jeffrey pines, the conifer needles lacquered by the late sun. Purple finch: a pair nesting in the ribs of the solar panels on the roof eave of a magazine office where I once worked, singing love songs to each other. The first time I saw a Canada jay I was hiking the Thunder Creek basin in North Cascades National Park when I came upon one in a stand of alder; the grove hushed with the moment. The first time I heard a Swainson's thrush was June 2022, in what I call the "old grove" in the woodsy park behind my place. The song—an ascending arpeggio—was sexy, like a robin had been gifted a wah-wah pedal. Paying attention, I've found, pays dividends measured in delights.

Maybe birds aren't your thing. Pick something else: wildflowers or trees or dragonflies or frogs. Like any natural phenomenon under the sun, those life forms can deliver revelations that will slot themselves into the mind's memory bank. Among those revelations will be a reminder of Earth's natural rhythms, the cycles within cycles that govern the living world.

Tan discovered something like that as the backyard birds invited her into a different pattern of time: "Spring Migration, Nesting Season, Fledgling Season, and Fall Migration." Another

California writer, the artist-essayist Jenny Odell, hit upon a similar discovery as she opened her eyes to birds "as an escape hatch from contemporary anxiety." In *How to Do Nothing*, which is a manual for resisting what she calls "the attention economy," Odell finds that when she noticed roses and crows, she also noticed different time scales. The crows in her urban neighborhood "don't see progress—they just see recurrence, day after day, week after week." Eventually she learns that "there were not only birds, but there was bird time."

Bird time is marked, most of all, by movement: migrations, peregrinations, calculated arrivals, scheduled disappearances. Bird time isn't just a different scale from modern human time—it's a different shape. That reshaping of time is among the most important reasons to be outside, as attentive and present as a songbird. To make lasting memories of the planet we've had and the places we're losing will require bending time into a shape that's easier to remember.

The mind-work of memory is mostly keeping track of time. This present moment is a cliff's edge on the great chasm of the past; the here-now is no thicker than an apple skin. Time, as much as attention, is the essential element of experience. Proust, naturally, understood this quite well. In a 1922 letter he wrote, "The image (as imperfect as it is) which seems to me best suited to convey the nature of that special sense [of memory] is that of a telescope, a telescope pointed at time."

Just as it's impossible to think about memory without considering time, it's also close to impossible to dwell on—or dwell *in*—more-than-human nature without thinking about time. Aged trees, ancient rocks, century-old tortoises, and venerable matriarch elephants are each, in their own way, a portal to the fourth dimension. To touch the bark of a three-hundred-foot-tall coast redwood is to make physical contact with the sweep of millennia. Time rests under the palm of your hand.

"Much of the beauty that moves us in the natural world is not the static visual splendor that can be captured in a picture, but time itself," Rebecca Solnit writes in *Orwell's Roses*. "The order that matters most is not spatial but temporal." I think it's fair to say that when many people accept nature's invitation to time travel, it's the grandeur of permanence that transports them away. The Grand Canyon's illustration of deep time impresses to the point of vertigo. A gnarled bristlecone pine, with its tree rings forming a living manuscript older than the Pyramids of Giza, transports us through history. I am lucky enough to have in the woods right behind my house a giant Douglas fir tree—about 120 feet high and a full 17½ feet in circumference—that, according to my calculations, is some 325 years old. It had already been living here for a century when, in 1792, British navigator George Vancouver encountered the Indigenous Coast Salish people who call this place home. When I stop and attune to the fir and attend to its age, the past collapses into the present.

For me, though, nature's most important lessons in time are less about permanence than *impermanence*—the ephemerality of every living thing. As much as I love the woods I walk daily, as much as I treasure the wildernesses I have visited, my most important personal ecology is the garden. I've spent nearly the entirety of my adult life gardening—including portions of it as a working farmer—and the garden has taught me many things. For starters, the importance of observation, since the gardener's attention is the best fertilizer. Also, an ethic of stewardship, because the planting and tending of an orchard is nothing less than a pact with the future. Most of all, the garden has given me an education in the contours of time.

My years in the vegetable patch have provided a hands-on tutorial in the endless rhythms of renewal and recurrence. Birth followed by death followed by rebirth and then repeated all over again and again, season by season, year after year. The garden mocks the notion of time as an arrow pointed toward the always-receding future. The garden reveals that the calendar is a map laid out as a circle.

Industrial society has always been ambivalent to antagonistic about the present's relationship to the past. In her book *Timefulness*, geologist Marcia Bjornerud warns that we moderns suffer from "a temporal illiteracy . . . a child-like disinterest in the past." Many preindustrial societies—Indigenous cultures, European peasant cultures—have thought about time in other ways. The ancient Greeks had two distinct conceptions of time. The best-known one—the time shape we inhabit today—is *chronos*, as in chronology. Chronos is the time in which things happen. In parallel to chronos exists *kairos*, a sort of time outside of time.

Kairos is usually understood to mean "the opportune time" or "the right time for the right thing to happen." The most capacious definition of kairos that I've come across is "time defined within a narrative." Whereas chronos is time marching forward like hashmarks on a yard stick—one damn thing after another—kairos is time invested with meaning.

The gardener and the naturalist can find a refuge in kairos, for nature is governed by opportune time. The wildflowers bloom when the time is right—no sooner and no later. The farmer plants when the time is right; the weather outside trumps the calendar indoors. In modern Greek, *kairos* is the word for "weather."

By rethinking the shape of the time, we can cultivate what Bjornerud calls timefulness—a "clear-eyed view of our place in Time." This is of the utmost importance when it comes to fighting shifting baseline syndrome. Resisting environmental amnesia requires a sense of time as much as a sense of place. We need to know not just where we are, but *when* we are.

The task is especially urgent given the hyper pace of environmental destruction in this age we inhabit. Scholars call it "the Great Acceleration."

The Great Acceleration describes the increasing speed and intensity of the human transformation of the planet since roughly the end of the Second World War. Since just 1945, the number of automobiles on the planet has grown from 50 million to well over 1 billion. The amount of plastic produced annually has grown from 1 million tons a year to more than 300 million tons a year. In only a few human generations, the total length of oil and gas pipelines globally has outstretched the total length of railroads. Time is a function of motion, and as human civilization moves faster than ever, time is speeding up.

The Great Acceleration is bad news for memory. It used to be that environmental change happened so steadily that it could be hard to notice—"the slow violence" of environmental destruction, in the words of one scholar. In the era of the Great Acceleration, environmental change happens so fast that it's hard to comprehend. Chronos doesn't help. Time configured as an arrow is an invitation to forget.

What might a memory-friendly shape of time look like? A circle is close to the mark, but a spiral is more like it—a helical form that recurs yet never exactly repeats. Envision a galaxy. Or, for a more earthbound metaphor, picture a snail's shell, a conch, the architecture of a succulent, the fractal arrangement of a romanesco cauliflower with its spirals within spirals. Memory feeds on patterns—and a spiral is pattern incarnate.

To combat environmental amnesia is not to resist all change, standing like a reactionary athwart history yelling "stop." No. This labor of remembering is, above all, about *noticing* the changes—in the hope that such noticing will lead to a clearer discernment about what we might save, what we have a chance of restoring. Spiral time assists in this by helping us recognize the patterns in nature and to be alert to when they change, to notice the jags and breaks and interruptions. Spiral time is a way of keeping the aberrant in sight, of not letting the new normal slip away along chronos's glide path.

A reforged shape of time? Maybe it all sounds impossibly mystical. If so, then consider this: To apprehend the recurrent renewals of the living world can be, at the very least, a way of remaining alert to nature's surprise beauty. It's a way of staying attuned to wonder, of not allowing yourself to be dulled even by beauty. Awe grooves the mind.

CHAPTER 2

BEAR WITNESS

In Bolinas, California, there's a little scrap of earth that neatly matches the definition of a leftover landscape or a hand-me-down habitat. The undeveloped lot on Kale Road is, in a way, remarkable for its plainness given that Bolinas has such postcard beauty: a seashore hamlet perched on the Pacific, with the broad-shouldered mass of Mount Tamalpais to the east and the silhouette of the Farallon Islands to the west. The lot is a shaggy, weedy place—mostly just blackberry brambles and poison oak, eucalyptus trees shedding their bark in scraggly peels. Someone's abandoned boat, covered in grime and flecked with moss, sits on the side of the dirt road. In the winter, the single path through the lot is a string of muddy puddles.

Ole Schell remembers it as something much more. Back in the 1980s, when he was growing up in Bolinas, monarch butterflies used to congregate around town by the tens of thousands and the spot on Kale Road was one of their favored hideaways. "There used to be endless butterflies just, you know, streaming off of branches," Ole told me. He recalls, with a bit of embarrassment, how he and his friends used to knock the monarchs off the trees and watch them fall to the ground by the hundreds. "We took them for granted."

Like many small-town kids, Ole left. He spent his high school years in San Francisco and after that moved to New York City. He got swept into the world of high fashion, lived for many years on Manhattan's Lower East Side, traveled the world making documentary films. The butterflies of his childhood were a long way away.

When he moved back to Bolinas in 2016, one of the first things he noticed was that the monarchs were gone. His first winter home, he was shocked to spot only one or two butterflies around town. The absence was glaring. But many of the locals—who had experienced the monarch diminishment as a steady decline instead of a sudden disappearance—seemed unalarmed. When Ole asked his mother about it, she said, "Yeah, there just aren't really monarchs anymore."

Ole offered me a ready-made—if flawed—metaphor to explain what had happened: "It's like the frog in hot water, right? . . . It's like things that seem normal at the beginning are no longer normal, or vice versa."

Perhaps his relationship with monarchs would have ended there. Then came the terrifying monarch population crash of winter 2020–21. Less than fifty monarchs were spotted at the historic overwintering sites in and around Bolinas. "It was more than a decline—it was a disappearance. They were in an extinction vortex," Ole said. At fifty-one, he thrums with the puckish energy of an avuncular hipster (tousled vibe, excellent jackets) who believes earnestness is cool. "It's not great to see a magical creature like that disappear." It felt to him like a nightmare in which you find yourself "naked and walking down the street."

He committed himself to doing whatever he could to bring them back. Thanks to some family luck, that was a lot. Ole grew up on a Bolinas cattle ranch that his dad co-founded with sustainable agriculture pioneer Bill Niman in the 1970s and which became one of the hot spots of the famed Bay Area food scene. Chez Panisse chef Alice Waters was a family friend; the Dalai Lama once paid a visit. Even though it hadn't hosted monarchs for years, the eucalyptus grove behind the shingled house his father had built from reclaimed wood was a known overwintering site. The property—an epic spot from which you can see San Francisco—had all the elements for a monarch home: the eucalyptus trees, plenty of open space, and, crucially, a year-round water source in the form of a cattle pond. Ole got the idea of turning a portion of the ranch into a monarch sanctuary.

He tore out sections of pasture and, with the help of a veterans' service group called Guardian Grange, got busy planting nearly two thousand nectar plants: California lilac, lupine, coyote bush, seaside daisy, flowering currant. He installed a fruit orchard whose blooms would also provide nectar. Weeks were spent building a deer fence and then rabbit-proofing the nectary.

"So this has basically turned into my life," he told me. "It's my full-time, nonpaying job."

Ole also turned his attention to the lot on Kale Street which, by the early 2020s, was one of only two butterfly overwintering sites remaining in Bolinas—this in a town where there used to be as many as twenty-nine. He convinced the landowner not to develop the site and worked with the utility PG&E to avoid wantonly cutting down the eucalyptus. He hustled donations to pay for the planting of quick-growing shrubs like wax myrtle to re-establish the kind of windbreak the monarchs would look for. He tacked a handwritten sign to a pine tree:

Monarch Winter Sanctuary
Quiet Zone
Stay on Trail
Share Only by Word of Mouth

I visited the Kale Road lot with Ole and Mia Monroe one sunny winter morning to see how things were coming along, having first visited the site a couple years before. Mia, you may remember, is the "monarch czar" who co-founded the Western Monarch Count back in the nineties. She had started working with the National Park Service when she was in college, and during a forty-five-year career at Muir Woods National Monument she had sparked a curiosity with nature among countless people. She was especially proud of her side gig with the Monarch Count, of having recruited so many acolytes like Ole, folks who had committed their time and energy to monitoring the monarch populations.

"Community science is a profound movement," she told me in one of our conversations. "It's the democratization of science. It recognizes that all of us are capable of being good observers." At the same time, she was all too aware of an awful irony at work: even as the number of volunteer monarch monitors had skyrocketed, the number of butterflies themselves had plummeted.

After the dismal 2020 crash, the estimated number of monarchs had looked pretty good for three winters in a row. The rebound—as modest as it was—had left the monarch watchers, in Mia's words, "ebullient." By the summer of 2024, however, it was clear that things were taking another grim turn. Very few monarchs had been spotted in their mating zones to the east, where record-breaking high temperatures were being blamed for either cooking the chrysalises or intensifying predation from wasps. By the time I made it back to the Kale Road lot in January 2025, the bad omens had been confirmed. The census numbers from the network of volunteer monitors were once again a line of zeroes. The biologists at the Xerces Society would eventually announce that the total figure for the season was a mere nine thousand monarchs statewide—a fraction of 1 percent of the historic high.

The scene at Kale Road was on-trend. Just a year earlier, Mia and Ole had counted a thousand butterflies there. But during the hour we spent at the lot, we spotted only two monarchs, one of them quite faded and tattered—evidence, Mia said, of an arduous migration.

They were trying to put on a brave face, but it was clear that they were down in the dumps. Ole described the situation as "catastrophic," "disheartening," and "dispiriting." Mia, decked out in a black hoodie with a monarch on the back and butterfly earrings, said, "We're all kind of depressed because there aren't monarchs."

But neither was ready to surrender. "I think you just have to have faith that they will come back, because of 2020 and those very low numbers, and they did come back," Ole offered. "What's the other option? Just to give up? That's a worse option."

"Even at this moment, it seems like a real seesaw, but you still want to be working towards offering a better future," said Mia, who was struggling with how to communicate the situation. "We need to all come together and cry together and then figure out how we'll talk about this. How do we prepare people and motivate them and allow them to process their emotions and their sadness?"

The monarch monitors are doing more than simply observing the natural world, coolly cataloging the butterfly decline on their clipboards. In gathering their annual census numbers, they are ringing an alarm bell, all in the hope that their unglamorous work might propel others to help safeguard the butterflies. Joined together in a community of care, they are bearing witness.

The contemporary, now-common meaning of "bear witness" is simply *to observe*. But that's a dilution of the original meaning, which comes to us by way of Christian conviction. Bear witness comes from a Greek word, *martyrion*, which essentially means *to make a statement*. To witness is merely to see; to bear witness is to declare what has been seen. While witnessing can be private, to bear witness is always a public act. In bearing witness, one carries the burden of testimony.

Bearing witness is one way of rejecting the lure of normalization and acceptance. Mia, for one, is very clear that the community science endeavor is a call to action, a cry of resistance. "People who bear witness don't just observe—they share and activate," Mia told me. "Bearing witness is a very powerful showing up and calling out what is happening."

This can take all sorts of forms. Bearing witness can be as small as a social media post and as big as standing up at a government hearing to speak one's piece. Bearing witness includes the yard sign, the bumper sticker, whatever flag you fly. It's the letter to the elected officials, the body in the street march, the participation in any public artwork—whether as artist or viewer—that transmutes beauty into testimonial. No matter what form bearing witness comes in, to turn a private observation into a public statement requires a measure of bravery. Many of us, after all, need to screw up some courage to share our beliefs with others.

Bearing witness involves putting emotions on the line. Emotion is imperative for another reason: it sparks memories. Big feelings, it turns out, create lasting remembrances.

"It is astounding how little the ordinary person notices butterflies," Vladimir Nabokov wrote in his memoir, *Speak, Memory*.

Nabokov is best known for being one of the preeminent literary stylists of the twentieth century, but he was also a lifelong, passionate lepidopterist who identified and named multiple butterfly species. Just like Mia, butterflies were part of Nabokov's emotional landscape from a very early age. "Few things indeed I have known in the way of emotion or appetite, ambition or achievement, that could surpass in richness or strength the excitement of entomological exploration."

Nabokov might have been right that most insects fly by anonymously, but the monarch butterfly is an exception to the rule. It does not go unnoticed, it is unique for the ways in which it arouses excitement, and—no matter its fate—it's unlikely to be forgotten anytime soon.

Danaus plexippus is the very emblem for an entire order of insects. It's the butterfly of street murals, of welcome signs at community gardens, of children's lunchboxes. Monarch butterflies are one of the few bugs (fireflies are the only other one I can think of) that have a whole tourism industry built around them. Monarchs are a gentle mascot of nature—the panda bear of insects.

Their fame is due, in part, to the fact that the monarch is big, bright, and unmistakable. With its pumpkin-orange wings embroidered with black lines and white spots at the wingtips, the monarch is a showy butterfly—"gaudy" in the words of one lepidopterist. Their wingspan can measure more than four inches across, nearly the size of some songbirds.

Mostly, though, its celebrity status comes from its incredible, continent-spanning migrations. There are two distinct populations. The western monarchs breed and lay their eggs in the intermountain West and then winter on the California coast. The eastern monarchs spend the summer in the Upper Midwest, New England, and as far north as Quebec and then travel to the mountains of Michoacán, Mexico, where they overwinter bundled together in groves of oyamel fir trees. Like high-flying birds, monarchs glide on thermal currents to make their long journeys. They have been seen nearly five thousand feet in the air, surfing the winds in the company of hawks.

Also like birds, monarchs use the planet's magnetic fields and the polarization of light to orient themselves and stay on course. They can tell the time of day and they know the day of the year, since they must understand the angle of the sun today as compared to yesterday to know exactly which direction to fly. Long before this science was confirmed, Nabokov intuited that butterflies can perceive "the pure element of time . . . time's common flow."

But unlike birds, no single monarch ever completes the entirety of their migrations. A monarch that returns to the winter groves her ancestors came from is the great-great-granddaughter of the monarch that laid the eggs she hatched from. How do they know where to go? It's a mystery. The whole journey is done by the instinct and intuition of a critter that weighs no more than a raisin.

Whether in the vast mobs in the Mexican fir trees or sprinkled across the eucalyptus and cypress groves of the California coast, the spectacle inspires rhapsodies. "Trees cloaked in the richest raiment imaginable," lepidopterist Robert Michael Pyle has written, the sky "filled with bright spinners and floaters and gliders, like doubloons falling in and out of focus." The phenomenon conjures metaphors. "Pendulous clusters masquerading as witchy tresses," Barbara Kingsolver writes, trees hung with "quivery, shifting, scaly black and orange bouquets." Mia has described them to me as "a poetic creature," said they connected her to "the heartbeat of nature." One monarch watcher called them "living stained glass—like rainbows without the association of rain." The ancient Greek word for butterfly was *psyche*.

All of this ignites uncommon devotion. One lepidopterist—who confessed to me that he's a bit tired that this one species gets so much attention—said that going to a meeting on monarch conservation was like being at "a crazy cat lady convention." Just as birders have their stories of their "spark bird," many of the monarch monitors wanted to share with me their monarch-love origin stories or their favorite moments with monarchs.

Mia had her childhood remembrances of meeting her Italian grandmother at the San Carlos train station, where the railroad tracks were lined with milkweed. Ole had his memories of the trees in Bolinas "streaming" with monarchs. Mary Dainton, a longtime docent at the monarch sanctuary in Pacific Grove, told me about what it was like when she arrived in California and discovered the butterflies.

Born in Wales in 1950, Dainton hadn't grown up with monarchs, but she was the sort of nature-loving child who kept rabbits, brought home frogs from the nearby pond, and enjoyed watching spiders hatch in the cupboard. Her first experience with monarchs came at the age of thirty, when she moved to the Monterey Peninsula to work as a nurse. She was enchanted from the start. "It was magical," she said. She recalls bicycling around town and seeing the monarchs wherever she looked. "You would be riding along Ocean View from Lovers Point all the way around to Asilomar, and the monarchs would be on the aloes, on the echium, flying on your face and landing on your hair. They were just everywhere."

Another veteran Monarch Sanctuary docent, a retired schoolteacher from SoCal named Stephanie Turcotte Edenholm, told me (twice!) about one of her most memorable moments from her dozen years of leading school groups at the site. "There was a school group of fourth graders from a local private school in Carmel Valley, and it was kind of drizzly and I was going to cancel. But they insisted they wanted to come. And there was a boy who was standing there when he had a monarch land on his shoulder. And it was going up his shoulder and up to his chin. I had all these kids standing around watching. And he stood there perfectly still and it rested on his check. And he stayed there for fifteen minutes until the monarch flew away. He had so much joy on his face. It brings tears to my eyes."

Scientists who study the brain describe two main types of memory. There's *semantic memory*. This is the stuff we know, like the dates of the American Revolution and a spouse's birthday and the name of that mom who's always at school pickup (what *is* her name?). Then there's *episodic memory*. Those are the memories of experiences, like, say, a frightening car accident or the day that never dawned because wildfire smoke blotted out the sun.

Stephanie's reminiscence of the boy with the butterfly is an episodic memory, in this case what neuroscientists call a "flashbulb memory." These are memories that are so expressive and sharp that they are like a scene from a movie. The classic example of a flashbulb memory is our recollections of major historical events. Where were you when you learned about the terrorist attacks of 9/11? What were you doing when you understood that COVID was going to change everything? Many flashbulb memories capture the peak experiences of our lives: a wedding, the birth of a child, the death of a parent. We remember what is meaningful, and nothing is more meaningful than such intimacies.

Emotions consume the mind's attention mostly as a function of physical arousal. When we enter a state of excitement—whether fear or anger, joy or happiness—we experience a rush of hormones—adrenaline or cortisol or dopamine. All of which explains why strong feelings make for vivid recollections. "Flashbulb memories are, on balance, more durable and accurate than most memories of day-to-day events," writes psychology professor Daniel Schacter in *Searching for Memory*. "The accuracy of memory is often directly related to the emotional arousal elicited by an experience, independent of whether it is positive or negative."

Memory is made of connections, and this is especially true of emotional memories, which activate several different regions of the brain. While memory is stored in the hippocampus, emotion is processed in the amygdala, an almond-shaped node that regulates mood. The encoding of a strong emotional experience involves a network of associations, a feedback loop between hippocampus and amygdala.

This is a dynamic process. An emotion is, quite literally, an act of motion. It's a stirring, an agitation. "I was moved," we say of some powerful experience. In that sense, our emotional memories aren't all that different from our embodied ones—both are physical. Descartes got it wrong. In fact, we *feel*, therefore we *think*.

As I talked to the monarch monitors across California, it became obvious that their commitment to the monarchs came, above all, from a profound emotional attachment to the butterflies. In a word, they cared.

Maybe that sounds corny. In our sardonic, irony-soaked culture, emotions are often an embarrassment. At the very least, this is true for the culture of much contemporary environmental advocacy. The language of modern environmentalism has become a technocratic argot of facts and figures and acronyms: parts per million of carbon dioxide, average global temperature increases, minimum viable population estimates for threatened species. Marshall the facts to make an argument and present the carefully collected evidence, and that will be enough to protect the environment. Or so the thinking goes.

But reason has its limits. We're a pretty smart species, and if we had been able to think ourselves out of the predicaments of global warming and biodiversity collapse, we would've done so by now. To have any real chance of avoiding the worst scenarios of ecological collapse will require something more durable than rationality alone. It will demand that the depth of our feelings matches the scale of destruction.

If it's true that some factions of the environmental movement have become unbearably cerebral, it's equally true that a species of love has always been at the heart of environmentalism. "It is not half so important to *know* as to *feel*," Rachel Carson wrote in *The Sense of Wonder*. "If facts are the seeds . . . then the emotions and the impressions of the senses are the fertile soil." In the sixty-plus years since those words were written—as the ecological crises have intensified—a growing number of voices have made the case for an emotive environmentalism. "This pain for our world is a measure of caring," the Buddhist eco-philosopher Joanna Macy wrote. In her book about the growing phenomenon of eco-anxiety, *Generation Dread*, Britt Wray argues that at this point in the Anthropocene heartbreak is rational: "It is logical to be horrified."

When we latch heart to landscape, we lay the emotional topsoil that can help cultivate a defense of the living world. Just like the embodied experience of being outside, emotions create bonds of memory strong enough to forestall forgetting. Feelings will prove one of the most effective antidotes to environmental amnesia.

But if emotional memories are sticky, they can also be sketchy. Even as emotions stamp experiences in the mind, they also smudge the memories of those experiences. Mood warps what we remember. This is especially true of episodic memory, which is infamously prone to error. Episodic memories aren't like a book or a photograph—a fixed record of the past. They are more like a story told and retold. They must be reconstructed each time they are retrieved, and with every reconstruction there's a chance for the introduction of errors.

At best, an episodic memory might be like a palimpsest—the original manuscript embellished with marginalia. At worst, an episodic memory is a photocopy of a photocopy of a photocopy—each replication becoming more faded and distant from the original. False memories are real things.

Forgetting and remembering aren't clear lines, like a border demarcating one nation from another. Nabokov knew this. In *Speak, Memory*, he writes about the "amnesic defects" of his original draft—"blank spots, blurry areas, domains of dimness." The mysteries of memory are reminiscent of the sprawl of the night sky. Even the sharpest memories can seem distant, a big star viewed through an unfocused telescope. Yet forgetfulness is also punctuated with bursts of clear sight, Venus and Mars shining out amid the ocean of dark matter.

If memory is essential to any effort to preserve and protect Earth, then the blurriness of remembrances makes for a serious challenge. How can we know that our memories aren't just daydreams of some concocted past? What if a skewed memory leads us astray even worse than forgetting?

One of the most seductive phrases in the English language is, "I can remember when." Do I hear a voice in the back of auditorium asking, with something of a sneer, whether this work of memory isn't just wallowing in nostalgia? *You shoulda seen the monarchs in my day, kid, back when the winters were really cold.* Every generation believes it lives after the fall.

Nostalgia is one hell of a drug, and it provides a cautionary example of how emotion can twist memory. Even though nostalgia is usually thought of as a bittersweet emotion—the ache for a lost time or place—it's also a balm, an experience that triggers reward centers in the brain. It's a longing that soothes, this autobiography dipped in rosy affect. When we pull an experience from past to present, we color it with the emotions of the here and now—a process that can lead us to misremember. The technical term is *mood-congruent retrieval.* Almost always, the past is viewed through rose-colored glasses.

Before nostalgia was a mood, it was a medical condition. The word was coined in 1688 by a Swiss medical student who was seeking to describe a debilitating homesickness he had observed in soldiers far from home. It comes from Greek roots: *nostos*, meaning homecoming, and *algos*, for suffering or grief. Originally, the term referred to a longing for place. For centuries, doctors believed it could be cured. The antidotes were the usual quack medicine of the time: leeches, opium, a visit to the Alps.

Today, nostalgia is usually meant to describe a yearning for a past more perfect than today. Along the way, it has taken on something of a shady reputation. "Nostalgia is to memory as kitsch is to art," a historian, Charles Maier, has written. Nostalgia has a reactionary feel. It's the political conservative's sour tonic, and for the better part of two centuries nostalgia has inebriated nationalists eager for greatness or conquest. I'm reminded of a sign I saw a few years ago on a road in rural Oregon. A fence was lined with *Make America Great Again* flags and then, at the driveway, a hand-painted declaration: "I miss the country I grew up in."

There's nothing automatically wrong with a fondness for the past; sentimentality is no sin. The problem with nostalgia is that it so often visualizes a mythic past, what the late historian Svetlana Boym calls "a romance with one's own fantasy." Nostalgia conjures the place or the time we *want* to remember—not necessarily the place or time as it truly existed. The fantasy is a mirage that turns out to be quicksand.

There's another emotion that is more likely to leave us with environmental memories that are honest and accurate. It's called *solastalgia*.

Like shifting baseline syndrome and the notion of the Anthropocene, solastalgia is a neologism of this bewildering age. The term was coined by an Australian environmental philosopher, Glenn Albrecht, around the turn of this century. It's a deliberate play on nostalgia: *sol* for solace and desolation, *algia* for pain. Albrecht defines it as "the pain or distress caused by the ongoing loss of solace and the sense of desolation connected to the present state of one's home territory." In the simplest terms, solastalgia is the homesickness one feels when, due to environmental destruction, home no longer feels like home.

Albrecht's portmanteau sprang from his own personal place on the planet—Australia's New South Wales, where he had borne witness to how the expansion of open-pit coal mining left the area's longtime farmers feeling dislocated even though they hadn't left their homes. But one can feel solastalgia anywhere. Solastalgia is the ache of seeing a once-vibrant forest turned into a charnel house of ash and dust. It's the agonizing confusion of watching the patterns of the seasons come unstuck from ancient rhythm: flowers blooming too early, migrating birds arriving too late.

When it comes to the work of ecological memory-making, the great virtue of solastalgia is that—in contrast to nostalgia—it doesn't crop the past. If it did, then it wouldn't click. For a place of solace to become a zone of anguish requires a fidelity to the reality of earlier times. Solastalgia also depends on specificity; anchored in intimacy, it counters abstraction. Butterflies aren't ideas—they are living, breathing beings, each with a will of their own.

Although I never heard any of the monarch monitors use Albrecht's fancy-sounding term, they all seemed to be suffering from a chronic case of solastalgia. Mary's foundational memories of her early years in Pacific Grove were of monarchs in massive numbers—the trees draped in them. "It was wonderful, it was magical." Those times are long gone. "You just don't see that now," she said to me, a Welsh lilt still in her voice after decades in the States. "It was horrible, it was really depressing."

Visitors to the Monarch Grove Sanctuary were also upset. Wearing her docent outfit—monarch pin on her fleece vest, monarch-festooned baseball cap—Mary would point to the educational signs that chart the monarchs' decline and do her best to explain the situation. "There were still people coming through, and they were really sad. You know, *Well, where are they?* They would almost be accusing us for there not being butterflies. People were mad."

The community's very identity was under threat. Tourists had been coming to Pacific Grove for more than a century to see the overwintering monarchs. By the 1980s, the city had built its civic image around the winged insect: "Butterfly Town, USA." The downtown has a Monarch Resort, a Monarch Pub & Restaurant, a Monarch Knitting. An orange-and-black monarch graces the city's official logo. All of it was coming to seem an anachronism. Pacific Grove's sense of self was draining away.

"It's so sad, because it's been a big part of my life from the time I moved here," Stephanie said. In 2012, she had published an illustrated children's book, *The Perfect Place,* about the monarchs' life cycle. The protagonist is a butterfly named Mozzie who makes her way from Spirit Lake in Idaho to the refuge in Pacific Grove—*the perfect place.* "Mozzie had never seen so many monarchs in one spot," the book reads. "First there were hundreds, and then there were thousands."

Now, Stephanie was beginning to question whether it would still be responsible to share the tale with schoolkids. "I always tell students that it is science-based fiction," she emailed me. "It never occurred to me that there would come a day when my book would be more fiction than fact. It makes me wonder if, one day, I will have to rewrite the book to align with what is happening."

Up in Bolinas, Ole was similarly bewildered. "When you leave a place, you come back, you think you're going to come back to how it always was," he told me while we were hanging out at the scrubby Kale Road lot. "So when I moved back here, I was surprised to see very few monarchs. It does make you feel a little bit like a foreigner in your own land."

This is hard stuff: anyone who has ever felt grief knows that even a hole has a presence. To hold loss in the mind requires what Nabokov called "a wing-stroke of the will . . . one of the bravest movements a human spirit can make." But courage-in-memory can make for a good grief, a productive grief that might alchemize emotion into action. That's Albrecht's hope. "Because its victims still hold within them a love of 'home,' the emotion of solastalgia also stimulates a response that opposes the forces that threaten one's sense of place," he has written. "Working with like-minded people to oppose the causes of the distress is cathartic."

Something as heavy as catharsis is needed to resist environmental amnesia for the simple reason that we are wired to experience shifting baseline syndrome. The mind is a prediction machine, always dividing expectation by reality to come up with the sum of normal. It's an ancient survival instinct: the only way we get through the day is by having a reasonable idea of what to expect. Our minds revert to the mean by forgetting the aberrant. "Forgetting is a form of learning too," neuroscientist turned environmental journalist Clayton Aldern writes in *The Weight of Nature*. "It would be irresponsible of the brain to pretend the world isn't changing. If we didn't forget the world of old, we wouldn't be able to adapt to the dynamic present."

Fair enough. There's no question that adaptation is imperative for survival on a planet changing faster than at any time in history. To navigate the storms, floods, and fires—the seasons knocked askew—will require adaptation strategies of the highest order.

But if adaptiveness is a virtue, blithe acceptance of reality is a vice. Acceptance is a form of surrender. On that awfully thin line between adaptation and acceptance lies our agency as moral beings. If that seems too fine a distinction, consider the history of Black Americans living under Jim Crow or Soviet Bloc dissidents living under communism. Survival required daily adaptation to cruel realities; acceptance of the systems of oppression would have been an admission of defeat.

Solastalgia inhabits that hairline divide between adaptation and acceptance. Solastalgia can prevent us from falling behind the veil of normalization. It serves as a check against accommodation thanks to the way in which it makes the world feel uncanny. The sheer strangeness of feeling homesick while still at home is a reminder that all this devastation may be many things—but it must never be considered normal.

The monarch butterfly probably won't go extinct. Although the species is endemic to North America, by now it lives many other places: Costa Rica, Hawaii, New Zealand. But the magical phenomenon of the monarchs' vast migration is in real danger of winking out. Reports of year-round monarch sightings around the Bay Area have led to conjecture that some populations may be staying put, perhaps in response to climate change. And what a loss that would be, for a choreography so long performed to disappear. A signature of home—erased.

Even now, something has been lost already: a sense of abundance. As recently as 2015, a paper by a group of eminent monarch researchers could describe the monarch as "undoubtedly the most well-known butterfly in the world" whose reputation was due to its "sheer abundance." The monarch used to be a common, backyard beauty; now it's an uncommon wonder.

Yet even amid the collapse, the monarch monitors hold on to a conviction that every monarch appearance can ignite passion in the public and sustain the efforts to conserve them. "If you've never had an experience of a thousand, then seeing five or four or one would still be wonderful," Stephanie insisted to me.

I got to see her perseverance one cool winter morning at the Monarch Grove Sanctuary. She and a colleague were doing the official New Year's Count, and Stephanie was sporting a black T-shirt with a bold monarch motif that she called her "lucky counting shirt." The sun had just cleared the hills to the east of town and the grove was empty of human visitors and butterflies alike. We spotted a scant few dozen monarchs on a eucalyptus branch swooping over the sanctuary fence from a neighbor's yard. I focused my binoculars to try to pick them out amid the long, slender leaves.

"They're a lot harder to see when they are so few," Stephanie admitted. "It's easier to count thousands than hundreds." She shrugged. "Thirty? Forty?"

Alongside the sanctuary's main path she found a grounder, a tiny monarch slowly flapping its wings in the dew-covered grass. "It's so small," she cooed as she bent down to inspect it from a careful distance.

We plodded through the grove of moss-swathed Monterey cypress, searching the overstory for clusters. There was nothing.

As we bushwacked back toward the sanctuary's entrance, visitors began trickling in. A mom came by with a boy toddling at her side. Stephanie told them she had discovered a monarch in the grass nearby. "I'll show you!"

"I read on the plaque that there used to be forty thousand of them," the mom said.

"Yeah, they are in decline," Stephanie offered.

"That is so sad," mom responded.

The boy skootched close to the monarch and took a long look. Mom turned to him and exclaimed, "Good job! You saw a butterfly!"

Really? Here was shifting baseline syndrome in action. As the world shrinks, our expectations shrink too. Ultimately, even awe gets downsized. It's all a matter of perspective. Extinction is unmistakable; there's no arguing with zero. But abundance, just like beauty, is in the eye of the beholder; one person's "few" might be another person's "many."

What would become of the monarch sanctuary in Pacific Grove if the migration were to completely collapse? What happens to a sanctuary when there's nothing left to safeguard?

Perhaps the preserve—along with the California state parks at Pismo Beach and Santa Cruz that are also dedicated to monarch overwintering—will become something like a memorial, a place where private grief is channeled into public mourning. When communities come together to mourn what has been lost, it makes for a potent form of bearing witness. To memorialize the dead and the gone—or to celebrate everything that remains but is in danger of disappearing—is one way of transforming witnessing it into the testimonial of bearing witness. Whether enacted through ritual or manifested through art, memorials can be a way of speaking truth to power, of insisting that one does not consent to the train of destruction. Memorials can be a statement that some losses are intolerable and inexcusable.

Such work is already happening around the world. About fifteen years ago, a group of British environmentalists began marking a "Remembrance Day for Lost Species" every November 30. The annual funereal gatherings now take place across the globe—Milwaukee, Sacramento, Australia, Slovenia. The Remembrance Day makes a point to honor the obscure species as well as the well-known: the endangered rusty patched bumble bee, the extinct Mariana fruit bat, the slender-billed curlew who is no more. When the Hawaiian bird Kaua'i 'ō'ō was declared extinct in 2023, staff at the Center for Biological Diversity held a wake for it and other lost species, complete with eulogies. As mourners read out tributes to each fallen animal, someone lit a prayer candle embossed with the species' image.

In Iceland, citizens came together in 2019 to grieve the passing of the Okjökull glacier, commonly referred to as Ok. It was the first glacier on the island to die, but it won't be the last, and for many Icelanders its disappearance came as a psychic blow, given how glaciers have always been a signature feature of the Nordic nation. "Where the glacier once touched the sky, there will be only sky," Icelandic poet and essayist Andri Snær Magnason wrote in his book *On Time and Water*. Icelandic environmental advocates and artists got the idea of erecting a plaque at Ok's former home, and they asked Magnason to write the text. It reads:

> OK IS THE FIRST ICELANDIC GLACIER TO LOSE ITS STATUS AS A GLACIER.
>
> IN THE NEXT 200 YEARS, ALL OUR GLACIERS ARE EXPECTED TO FOLLOW THE SAME PATH.
>
> THIS MONUMENT ACKNOWLEDGES THAT WE KNOW WHAT IS HAPPENING AND WHAT NEEDS TO BE DONE.
>
> ONLY YOU KNOW IF WE DID IT.

The tombstone is an act of bearing witness. It's a challenge to the present delivered in the form of a message to the future.

Another powerful example of environmental memorial-making comes from Maya Lin, the artist best known for designing the Vietnam Veterans Memorial in Washington, DC. Her latest—and, she says, final—memorial is a multisite, multiyear project called "What Is Missing?" One of the project's installations was titled "Ghost Forest." The exhibit involved transplanting forty-nine dead Atlantic white cedars, killed by climate change, to Madison Square Park in Manhattan, where they stood in an arboreal graveyard for the better part of a year. "Each tree," Lin said at the time, "is a monument to climate change."

"What Is Missing?" is a direct response to the problem of shifting baselines. "Over time we forget how abundant the natural world once was," Lin declares in an artist statement and then asks, "How can we protect it if we don't even realize it's missing?" To help reveal what's gone missing, Lin has made the project interactive and asks museumgoers, exhibit viewers, and visitors to her website—whatismissing.org—to share their ecological memories. The offerings are a record of a world that is withering: a "lack of abundance of sea life" off the coast of Florida, tree frogs in California that "seemed to disappear," a field of sunflowers on the Canadian prairie "razed for the construction of more houses." A woman from Oregon City submitted this memory: "I remember countless monarch butterflies fluttering around my farm in the summertime. Now it's exciting when I see just one, which is not very often."

We have an obligation to honor the dead. But we also need to be careful not to drown in requiems—if for no other reason than the fact that grief can overwhelm memory. Neuroscience has shown that stress, depression, and anxiety can, over time, impair the workings of memory. Under the weight of too much sadness, the mind's executive functions seize up, as does long-term memory encoding. Despair is an enemy of memory.

Joy, in contrast, energizes memory. The mechanics of arousal and long-term memory formation are the same for sadness and happiness alike, but not all emotional effects are created equal. Brain researchers report that positive emotions have a distinct effect on memory and lead to enhanced associative memories. Positive emotions simply have a different valence from negative ones. Partially this is a matter of positivity bias, partially it's a matter of reward-seeking. Happy memories feel good, so we hold on to them tightly.

If we're to overcome environmental amnesia, our memories of the natural world can't only be elegies. The strongest and longest-lasting memories of nature are likely to come in the shape of ecstasies.

Poetic souls figured this out independently of the MDs and PhDs. "What right have I to grieve, who have not ceased to wonder?" Thoreau wrote a long time ago. Much more recently, *New York Times* columnist Margaret Renkl reminded readers: "The fallen world . . . is not the only world. We also dwell in Eden, and every morning the world is trying to renew itself again. Why should we not glory in it, too?"

One of the most impassioned cases for joy I have heard comes from the English nature writer Michael McCarthy. In his book *The Moth Snowstorm*, he celebrates joy as the fulcrum with which to raise an ethic of care for the earth. Joy, he declares, "can be made the basis of our defence of the natural world in the terrible century to come." As it happens, McCarthy's love affair with the living world was sparked by a childhood encounter with backyard butterflies, the insects like "jewels flashing dazzling colour combinations." Butterflies flew into his soul, and they never flew out.

McCarthy, born near Liverpool in 1947, is a baby boomer, "the generation who, over the long course of our lives, saw the shadow fall across the face of the earth." He calls this *The Great Thinning*. "Abundance, blessed, unregarded abundance, has been destroyed." Among the losses was the disappearance of "the moth snowstorm." In the England of the 1950s, it was a common experience to drive at night through a blizzard of insects lit up by the headlights. McCarthy found that when, in the twenty-first century, he mentioned the moth snowstorm to other Britons his age, they responded with great emotion as the memory unlocked a corner of their minds.

The moth snowstorm is similar to what's known as the "windshield phenomenon." People of a certain age—including Gen Xers like me—are old enough to remember when a road trip would leave the front of the car splattered with dead bugs. When I was a kid, the bodies of insects would cover our family's blue Buick station wagon anytime we went on a road trip. I recall gas-station stops spent picking off broken wings, busted carapaces, smudges of guts smeared across chrome and glass. That rarely happens anymore. The invertebrate die offs of recent decades—the insect apocalypse—has left our windshields all too clean. "A child born in the generations after mine can never know" the moth snowstorm, McCarthy moans.

Even amid all the losses he has borne witness to, McCarthy is insistent about the power of joy—joy not just as fleeting experience, but joy "as a conscious, engaged act of defence" of Earth. A "sudden intense love stemming from an apprehension that there is something extraordinary and exceptional about nature as a whole."

Reminds me, once more, of Nabokov. "The highest enjoyment of timelessness . . . is when I stand among rare butterflies and their food plants," he wrote in *Speak, Memory*. "This is ecstasy, and behind the ecstasy is something else, which is hard to explain. It is like a momentary vacuum into which rushes all that I love. A sense of oneness with sun and stone."

I felt some of this myself when I went monarch-counting with Mia and Ole on a late-autumn morning during one of those "rebound" years when the monarch populations were looking solid. We had started at Ole's weedy spot on Kale Road, but it turned out to be a bust. We spotted just a single monarch after nearly an hour, so Mia suggested we check out a nearby grove where just a week earlier a professional lepidopterist had counted nearly four hundred butterflies.

It was another leftover landscape—a scrappy shelterbelt of eucalyptus on the edge of the Bolinas's wastewater treatment facility. As Ole shared memories of riding his motorcycle there as a kid, Mia started to scan the canopy. A local guy with his dog on a leash passed by, head down and earbuds in. *It is astounding how little the ordinary person notices butterflies.*

Mia was the first to see the monarchs. There were three, then four. "I just saw five," she said.

We spotted the cluster. Ole went wild. "Look up!" he shouted. "Ha ha ha! Look at this! Look at this!"

We angled for a closer look at the clump of wings swathed around a eucalyptus branch. There they were: ochre strands hanging earthward, leaf after leaf shingled in copper.

"And look, there's a bunch up here," Ole said. "Oh, look at the tree right there. On the top—the whole tree." He was laughing madly, laughing irrepressibly and shouting with glee. "The whole tree! The whole tree!"

Next, something even more amazing happened. Nearly the entire cluster took to the air together, all at once. "This is what I call *the orange cloud*," Mia said.

Monarchs can't fly at air temperatures below fifty-five degrees. As it gets warmer than that, they may be roused from their torpor in an instant. And that's what was occurring now. Some slight change in temperature, or subtle shift in air pressure, or the sudden escape of the sun from the grasp of the morning fog had sent them exploding into the blue.

Orange cloud was apt enough. The eruption of color sparked another phrase in my mind: *breath of butterflies*. This was the living world's exhalation, made visible in a burst of color.

Mia got busy counting the butterflies. There were about 170, she estimated. Ole sprawled out in the grass, flat on his back gazing skyward at the spectacle. A mammoth monarch flapped toward my face, seemingly as big as a bird.

As we were leaving the site, Mia turned to me and said, "It's what you imagine heaven is like: being in a meadow full of your friends, with monarchs flying all around."

It was unforgettable.

CHAPTER 3

MAKE A RECORD

When the ice came to cover Wisconsin's Pewaukee Lake the first winter that Gay Ruby and her family lived there, she instinctively wrote down the date: December 12. Gay had grown up in the suburbs of Milwaukee, so noting the ice-on date seemed natural to her. Lots of people do that sort of thing in Wisconsin, where winter ice fishing is a beloved tradition and where it's often a competition to see who can get their dock in the water first for summertime boating. Gay had no formal scientific training, but she had always been a "weather geek." One of the reasons that she and her husband, Jake, bought a lakefront property was so that she could have a place to watch the weather unfold—the storms rolling up, the lightning strikes over the water. "I love nature and I love watching sunrises and sunsets and storms," she told me. "When it froze the first season we lived here, I needed to record that. So I did it."

The Rubys bought their home in 1986, a "truly beat-up house" that they tenderly restored, and for the next decade her careful chronicle was merely a passion project. On three-by-five lined index cards Gay jotted down the ice-on date every winter and the thaw date every spring. The ice records weren't something that took a lot of time, but as the years went by the exercise began to feel important. The habit gave her a window into the natural world. "This has made me more aware," she said.

Then, in the mid-nineties, she read a story in a local newspaper, the *Lake Country Reporter*, about a woman on nearby Nagawicka Lake who was making similar records and sending her data to a researcher, Kenton Stewart at the University of Buffalo. Gay thought she'd like to be involved too. "Oh, I should be doing that," she said to herself. "If I can be a citizen scientist, I'm all in."

She was soon a member of Professor Stewart's Community Lake Ice Collaboration, a network of volunteers across the northern United States and Canada who track the ice-on and ice-off dates of their local lakes. Stewart (who passed away in 2023) was a limnologist—a scientist who studies lakes and other freshwater bodies. He had been trained at the University of Wisconsin, which has one of the country's longest running records of ice dates; the statistics for Madison's Lake Mendota stretch back to 1852. Stewart knew that to best track change across time would require a breadth of data as well as a depth of data. In the early 1980s, he began to enlist ordinary people to collect ice dates from as many northern lakes as possible.

This was the pre-internet era, so, to recruit lake watchers,

Stewart scoured phone books to locate the numbers for local marinas, chambers of commerce, and state conservation offices. Sometimes, he picked up new volunteers by chatting with curious passersby he encountered while doing water sampling. Some volunteers came via personal connections; Gay eventually recruited her parents to monitor the lake closest to their house.

Stewart was unwavering about staying in touch with his volunteers. He called many people annually, and sent all of them regular postcards or self-addressed stamped envelopes so that they could easily send him their data. The clockwork-like winter and spring letters came with individualized notes asking about the broader health of people's local lakes ("Are there still some zebra mussels?") or offering his own weather pleasantries ("Spring can be a wonderful time of year"). Gay told me, "I can't assume I was the only one he did this to. He was trying to make sure people responded. He cared deeply, which made me care deeply."

Pewaukee Lake is located about twenty-five miles west of Lake Michigan, around where the standard-issue office parks of exurban Milwaukee surrender to the bur oak, red maple, and birch of the rolling prairie. It's a big lake—some 2,500 acres of blue-green waters that kick up whitecaps when the wind is gusting. On a map it resembles a dachshund-shaped jigsaw puzzle piece. By the time Gay's records were entering their third decade, she had become something of an authority about the place.

"I am the little local expert," she said. "I am known as the ice-on and ice-off person." Once, at a "ladies' luncheon," someone casually wondered when the lake ice would break up and

Gay—who before retiring had a career as a journalist and later as a public affairs officer for the Federal Emergency Management Agency—rattled off a string of dates and yearly averages. "Every single woman was looking at me like, *Who is this weirdo?*"

In 2007, an area newspaper published a front-page story based on her observations with the headline, "Climate Warming Could Alter Lake Life." In the early 1960s, the story reported, *Life* magazine had dubbed Pewaukee Lake the "ice-boat capital of the world." But by the late aughts it was clear that ice season was getting shorter. Gay had the figures to prove it. In the first ten years of her recordkeeping, the average length of the ice season at Pewaukee Lake was 101 days. When the newspaper called in 2007, that number had decreased to 95 days. Winter was shrinking.

The article included a quote from a professor at UW-Madison named John Magnuson who told the newspaper, "Personally, the loss of ice is kind of sad." Like Kenton Stewart, Magnuson was an important figure in limnology. He was an early promoter of what's called "long-term ecological research," which is the idea that to really understand ecosystem change requires watching a single place over a long period of time.

"It is the unusual person who senses with any precision changes occurring over decades," Magnuson wrote in a paper published in 1990. "At this time scale, we are inclined to think of the world as static, and we typically underestimate the degree of change that does occur. Because we are unable directly to sense slow changes and because we are even more limited in our abilities to interpret cause-and-effect relations, processes acting over decades are hidden and reside in what I call 'the invisible present.'"

Like the concept of shifting baseline syndrome, the invisible present puts words to the often-ineffable feeling of trying to understand change swirling by. One's experience of time—much like one's definition of abundance—is a matter of perspective. "The field cannot be well seen from within the field," Ralph Waldo Emerson once wrote. For many of us, most of history beyond yesterday is flat.

Long-term ecological research (or LTER) is an important antidote to such flattening. It puts the present into context and, in doing so, helps scientists to disentangle the signal of unexpected environmental change from the noise of constant change. The research done at National Science Foundation–funded LTER sites like the HJ Andrews Experimental Forest in Oregon or the Harvard Forest in Massachusetts or the Beaufort Lagoons in Alaska is invaluable. But there's a hitch: most of that work is done by professional scientists.

The lake monitoring is different. What's so cool about this long-running effort is how accessible it is. Butterfly counting requires a pair of binoculars, a bit of training, and some practice to learn how to count a clump of insects thirty feet high in a tree. All you need to chart the ice on a lake is a pair of eyes, a calendar, and a pen.

The tracking of the seasons—*phenology* is the formal word—is perhaps the most democratic form of citizen science. To date, the Community Lake Ice Collaboration (which continues today under Sapna Sharma at York University in Toronto) has recorded ice observations for more than one thousand lakes in North America. The project has collected more than 52,000 observations and counting. Nearly a thousand individual volunteer monitors have been involved.

The lake watchers are counteracting environmental amnesia in real time. They're showing how we can each—in our own way and in our own place—become ecological memoirists. The lake watchers' steady vigil is contributing to a larger body of shared knowledge—the creation of collective memory. They are writing the first drafts of natural history.

If you want to remember something, write it down. That wisdom is as old as the first writing instrument and to-do list. Even the sharpest episodic memories aren't immune to the erasures of shifting baseline syndrome—attentiveness and sensitivity only go so far. Plus, we all suffer from self-doubt. Maybe it was always this warm at Thanksgiving? Maybe the summers were always so dry? The brain remembers by constantly rewriting itself, but if a culture is to remember it must resist the temptation to rewrite. To truly combat ecological amnesia involves the collection and organization of information that can be placed into the vault of history.

So, then, the next element in the antidote to shifting baseline syndrome: Make a record.

Inevitably, much of the labor of charting transformations in the living world is done by professional researchers. But just as anywhere is someplace, anyone can be a community scientist. When you commit to making some kind of chronicle of the environment you live within, you begin to write a record that can pass through the folds of time.

Some of this is self-evident. To sit in the backyard and write an observation of the flowers in bloom and the birds who are calling is to make a record that transcends that singular moment. The present is secured fast by words on the page, characters carried forward into the future. Record-making endows perception with precision.

Note-taking is a mnemonic prosthesis, a way of extending the reach of memory. And the simple act of note-taking helps the mind to form memories that can outlast the record itself. Even if the notebook ends up lost, you may remember the content of the notes. Teachers encourage students to take notes for good reason: the mere action facilitates learning.

Here's how. The brain transforms a short-term memory into a long-term memory via a process called encoding. In the encoding mode, the mind is getting ready to make a memory. "Without encoding," research neurologist Richard Restak reports, "recall is impossible." But encoding is more than merely paying attention. It's a process—still not fully understood by scientists—in which the mind is paying attention to encoding itself, to the act of memory-making.

Encoding occurs most readily when the hippocampus—that seat of memory—is in a state of arousal. Neuronal activity in the hippocampus prior to the reception of new information allows us to learn the information more effectively. Exercise and heightened emotions both stimulate the hippocampus. But what if we're not experiencing an emotional or physical high? How can we spur our minds into encoding mode in more routine situations?

Simply having pen or pencil poised over paper is one way to create lasting memories. Recent research has demonstrated that note-taking via handwriting is superior to taking notes with a keyboard or even a stylus and a tablet device. According to one study, research participants who used a paper notebook had stronger memory encoding and retrieval than those who used electronic devices. Their note-taking was faster and, significantly, more accurate. Another investigation found that writing by hand—in comparison to typewriting—involves widespread connectivity across a range of brain regions. This is due, in part, to coordination across conceptual systems, visual systems, and motor systems. As any knot-tying sailor or guitar player knows, the hands remember.

Spatial cognition is also at play here, which brings us back to the embodied nature of strong memories. We are constantly orienting ourselves in time and space, and when we write on a piece of paper, we're doing similar orientation. The scrawl on paper engages both episodic memory and spatial memory as our mental cartographer makes a map of the page. A notebook scrap is its own kind of memory palace.

I've done this myself—anyone can.

A couple of years ago, I started keeping a nature-and-garden diary. It's nothing much. I copy the day's high and low temperatures as well as any precipitation and then write a sketch of the weather: "February 12—A drizzle turning to steady rain and back to drizzle." I jot down the happenings in the nearby woods: "May 20—New plants trailside: fringecups and waterleaf. The maples completely leafed out." I note any interesting bird sightings: "August 21—A pair of ospreys over the lake." I track the comings and goings in the vegetable garden: "April 10—three lbs. of cosmic gold potatoes planted . . . May 15—Harvested Chinese cabbage and turnips."

One of the great values of my kitchen-counter recordkeeping is the way in which it allows me to track changes from year to year. My neighbors tell me that the daffodils here in Upper Cascadia used to pop in mid-to-late March; now it's early March. Thanks to my nature-and-garden journal, this is more than a hunch. I have the receipts to prove it. Ink on the page is a check against the perceptive mush of shifting baseline syndrome.

The habit of recordkeeping also helps me to stay attentive. I find myself noticing events in the woods because I know I'll want to write them down later. And when I go back and read the entries from months past, I find that the scribblings ("April 23—A steady, soaking mist all day . . . many Pacific wrens on the Salal Trail . . . First harvest of the season! Lettuces.") end up triggering other memories, opening up another portal of time.

In the course of my phenological recordkeeping, I've come to believe that such workaday observation has another benefit: it can give a measure of shape to the disarray of a world unraveling.

It's easy to think of global warming as a steady, linear increase in planetary temperatures. And it is that—the Earth's fever rising decade by decade. But the *effects* of climate change—extreme storms, unprecedented wildfires, historic floods—arrive in a nonlinear fashion. This is why some people prefer to use the term *climate chaos* instead of climate change. Many of the impacts of global warming are what scientists call stochastic—random and hard to predict. While one corner of the world might be experiencing its hottest temperatures and worst drought in memory, another might be suffering from its worst floods on record.

In some ways, chaos can be unforgettable. Peak moments like disastrous hurricanes or punishing heat waves will be rendered as flashbulb memories. No one who experienced the apocalyptic Northern California smoke storm of 2020 will forget it: the day the sun failed to rise was too terrible to be erased from the mind.

But, in general, chaos is confusing. When chaos is the status quo, the brain's penchant for pattern detection stumbles. It becomes harder to locate and hold on to the associations that are the keystone of knowledge. Climate change, University of Cambridge geographer Mike Hulme has written, "disorients our memories."

The autumnal changing of the leaves is a good example. It used to be that along most of the East Coast, "peak color" would reliably occur between October 10 and October 20. If leaf change were simply happening later in the fall, it might be easy to track. Instead, the dates are all over the calendar; the rate of variability in peak color dates has doubled in less than a decade. It's a random jumble, which is what makes it difficult to recall year from year.

Another example comes from Southern California, which is now experiencing what has been dubbed "hydroclimate whiplash." During the twenty-teens, the regions suffered from a string of drought years that, among other impacts, killed tens of millions of trees in the southern Sierra Nevada. Then came record-breaking precipitation in 2022–23, which was followed by another exceptionally wet winter the next year. And after that? The spring and summer of 2024 was one of the driest ever recorded. Under hydroclimate whiplash, the dry years are getting drier while the rainy years become rainier. Such dramatic swings are a recipe for disaster, as the January 2025 fires in Los Angeles tragically demonstrated. The wet years laid the fuel in the form of abundant grasses, and then the dry year set it all aflame.

Climate whiplash proves that baselines don't always shift along a graph's steady plotline; sometimes, the shift is more like a squiggle. The mind's instinct for pattern-seeking shipwrecks against the erratic. Once again, velocity confounds. It's not just that the climate changes are nonlinear—they are also abrupt and accelerating, which makes them difficult to grasp. As climatologist Andrew Dessler has written, "The speed of us passing limits is mind bending."

My little phenological journal isn't going to counteract global warming. But I have found that it's a counterbalance to the disorientation of climate chaos. To mark and measure the seasons, to note the daily happenings of the living world, helps me to see patterns more clearly. And in perceiving patterns I then notice the abnormalities too. For me, the habit of keeping a phenological journal is a deltaic process—the observations and notations piling up in a great alluvial fan, spreading outward and growing deeper. Over time, the chronicle comes to take on a life all its own.

The lake-ice figures kept by Dennis and Joy Krubsack go back decades longer than those of Gay Ruby—all the way back to 1953, when Dennis's parents bought a summer cottage on Pine Lake. Dennis's father had a small furniture manufacturing business in nearby Clintonville, Wisconsin, that Dennis eventually took over. In 1973, he and Joy bought a house on the north shore of the lake, a split-level home in a dense stand of candle-like second-growth white pine. From their living room they enjoy an unbroken view of the water—a perfect plate of sapphire necklaced with a mix of year-round residences and getaway cabins.

Pine Lake is located in the midst of Wisconsin dairy country, a landscape quilted with grain fields and pastures, the red-washed barns like something out of a children's picture book. Pine Lake is a scant two hundred acres, no more than forty feet at its deepest, the last lake in a three-lake string that starts as spring-fed waters and ends in a beaver-engineered marsh. In the winter it (usually) freezes thick and early. Back in the eighties, Dennis remembers, Firestone used Pine Lake as a testing ground for its winter tires. Professional drivers spent weeks on the lake making donuts in big trucks and spinning out across the ice surface.

The Krubsacks love the place and have committed themselves to stewarding the lake. "If you can just take care of your own little area, and care for it, and do what you think is the best for it—it helps," Joy, a retired middle school science teacher, told me when I paid them a visit on a too-warm spring day. Not long after they moved to the lake, they began taking water clarity and oxygen samples for the Wisconsin Department of Natural Resources. They worked with neighbors to put the lake's Gibson Island into a conservation easement. For a time, Dennis served on the board of the Cloverleaf Lakes Protective Association, one of the oldest of its kind in the state.

And that's how limnologist Kenton Stewart found them, sometime in the 1990s. They were able to give Professor Stewart their own family records of freeze and thaw dates, along with ice-off dates going back to 1940 that Dennis's parents had picked up from an area old-timer. For decades, they dutifully mailed in their dates, and often talked to Stewart on the phone, who knew the area well and liked to chat about invasive zebra mussels and an aquatic weed called Eurasian watermilfoil.

The Krubsacks told me that the decades-long habit of lake-ice monitoring has allowed them to notice things that others miss. The idea of shifting baseline syndrome wasn't something they had heard of, but they were certain they had seen it. "I think a lot of people say they see changes, but we *know* it, because we have the data," Joy said. "They know what they see, and that's all they know. And every generation and every person starts a new starting point, and they don't understand what they don't see. And that creates an amnesia."

But when I asked them about their reasons for doing the lake monitoring, they at first struggled to put words to their commitment. "We're just crazy people, I guess," Dennis said with a laugh. Gay Ruby also had a hard time with the question. "You had asked me why I do this, and I really had to kind of think about it," she told me.

The monarch monitors are driven by the flame of urgency. Their act of bearing witnessing, they feel, is central to preventing the extinction of butterflies. Lake watching is more of a low-key vigil. And yet what the lake watchers seem to lack in intensity is balanced by a depth of commitment—passion proved by perseverance. Over time, the longevity of their records has taken on a gravitational force. Once they started keeping the data—and after they kept at it decade after decade—they felt a responsibility to keep it going. "There is value in the record, and you don't want to drop it and have it lost," Joy said. "We have a sense of obligation to the record."

I heard something similar from Pat Collins, a retired Wisconsin Department of Natural Resources staffer who for thirty-three years has been monitoring Perch Lake, which he describes as "a little glacial lake—deep, cold, clear." Like the other lake watchers, Collins initially started collecting lake-ice data "just for fun." For years, he kept his tally of ice-on and ice-off dates scribbled on the inside of his garage door. After he hooked up with the Community Lake Ice Collaborative, he began to feel a duty to something larger.

"I like the phrase citizen scientist—that's kind of how I see myself now—trying to get good information out there to people," he told me. "It feels like a community even though you don't know the other members. But you know that people are collecting this information and sending it in. It feels like you're doing something collectively."

Gay was emphatic on this point. "This tiny bit that I'm doing along with the hundreds of other people who are also doing that—ice conditions are a small indicator of climate change," she said. "It's not something that I spend a lot of time on, but it's meaningful. It does matter."

She is so determined to keep the work going that, now in her seventies, she's come up with a succession plan. If she's out of town near the freeze or thaw windows, she relies on a neighbor to do the monitoring; she's now grooming that person, a friend of her daughter's, to eventually take it over. The Krubsacks also have a network of understudies who they are hoping will take over their records someday. "I'm watching and scouting who can take our place," Joy said.

"I can't imagine not doing it," Gay told me. "It's in my marrow, my bone marrow."

Can such homespun chronicles really counteract shifting baseline syndrome?

Yes, actually. There's a whole academic discipline called historical ecology that is focused on studying the interactions of humans and wild nature over long periods of time. Historical ecologists rely on a wide range of sources to try to tease out the causes and consequences of humans' influences on the more-than-human. Source material for historical ecology include the hieroglyphics on Pharaonic monuments, old paintings, property deeds, land surveys, insurance records, antique maps—the relics of daily business, made meaningful by the passage of time. Regardless of whether it originally had an ecological purpose in its own time—and usually it didn't—such historical ephemera can help prevent environmental changes from fading into forgetting.

One of the more famous examples of historical ecology is Henry David Thoreau's massive journal. In his published writings and speeches, Thoreau presented himself as a woodland flaneur, rambling about the forest plucking huckleberries. But every ramble had a mission—to observe the natural world closely. His private journals reveal another side of him: a type A obsessive meticulously logging the life cycles of hundreds of species and constructing his own private taxonomies of snow and ice. I have a set of Thoreau's journals, and the botanical index alone runs to more than two hundred pages.

Pick a species at random from the list: skunk cabbage, a flowering perennial that thrives in wetlands. It enjoys more than eighty mentions in Thoreau's journals. On April 2, 1856, he recorded, "the skunk cabbage has been in bloom some time and been frost bitten and decayed. . . . I find no other flower nearly so forward as this." And so on, with the focus of a PhD biologist at an official long-term ecological research site. No wonder the massive record has served as a useful baseline for contemporary researchers. Thoreau's journals make all too clear that spring at Walden Pond ain't what it used to be. These days, skunk cabbage in Massachusetts normally blooms at the beginning of March and, in some years, as early as February.

One of the most haunting pieces of historical ecology I know of is the "hunger stones" that appeared on the banks of the Elbe River during summer 2022, when the river's levels dropped devastatingly low amid one of worst European droughts in living memory. The stones marked the nadir of droughts from centuries past. A hunger stone near the Czech town of Děčín included the dates of earlier droughts—1417, 1473, 1616, 1707—and was inscribed, "If you see me, then weep." One stone, near Bleckede in Germany, carried an inscription that read, "When this goes under, life will become more colorful again." Like the tombstone for Iceland's Ok glacier, the stones are a record of the present conveyed as a message to the future, a warning cast across the centuries.

Whalers' records have proven to be an especially valuable historical resource. For the better part of two centuries, whalers sailed to every corner of the ocean. On those journeys they kept diligent records of ocean conditions, rainfall, and the direction and speed of the wind. They made notations about sea life and the birds they encountered. Together, whalers' records form one of history's largest surveys of ocean conditions.

Beginning in 2019, the Woods Hole Oceanographic Institution began to digitize them. So far, they've dug into more than a hundred whaler logbooks and in the process have catalogued thousands of weather records. NOAA is doing something similar with logbooks from the US Navy. Volunteers with NOAA's Old Weather project have digitally transcribed some 14 million past weather observations from naval records going back as far as 1849.

Sometimes natural history can be found on the menu. When researchers at the University of British Columbia looked at Vancouver restaurant menus dating back to the 1880s, they confirmed a transformation of marine life. A seafood restaurant's selection is an expression of what's in the sea—and the sea is changing. Vancouver restaurants used to serve a lot of salmon; now they serve more squid. In response to the findings, the former executive chef at Hotel Vancouver spoke movingly of the changes he's witnessed just in the last generation: "You get to know and love certain items, and when they decline and you no longer see the same fish around, it is a bit sad, because you just have such fond memories."

One last example. In the Kiso Mountains of Central Japan, Shinto priests for centuries have been carefully watching the waters of Lake Suwa. When the lake freezes, temperature changes from day to night cause the ice to expand and contract. This cracks the frozen lake surface and creates a ridge, called *omiwatari* in Japanese. According to Shinto belief, the *omiwatari* is formed by the footsteps of a male god crossing the frozen water to visit the shrine of a female god on the other side of the lake. This is considered an important event by the Shinto priests, and it has been recorded by at least fifteen generations of priests going back to 1443.

The lake ice records at Lake Suwa are a couple of centuries older than the steam engine, and they convincingly show the changes that have happened during that time. In the first 250 years of the ceremonial recordkeeping, there were only three winters in which Lake Suwa didn't freeze over. Between 2005 and 2014 alone, that has occurred five times.

These kinds of historical records fit perfectly into Daniel Pauly's original notions about how to counter shifting baseline syndrome. In his seminal essay, Pauly floated the idea that "ancient observations" and "historical reflections" offer a way to correct collective amnesia. "The big changes happened a long time ago, and all that we have as evidence are anecdotes," he wrote. The whalers' logbooks and the Vancouver menus are anecdotes that make the invisible present apparent.

Academics in the field of memory studies have a technical term for shared stories held in common: collective memory. A plainer word is *history*. History is memory practiced together. The best—really, the *essential*—antidote to environmental amnesia across generations is the preservation of history. The writing of natural history—every bit as much as petitions and protests, lawsuits and marches and grassroots organizing—is an invaluable tool for resisting the devastation of the planet's life systems.

In the realm of human affairs, it's well understood that forgetfulness is an ally to the forces of oppression. The despot, the dictator, and the imperialist have always understood the political uses of what Czech dissident writer Milan Kundera called "organized forgetting." Throughout the centuries, historical erasure has been a potent tool of authoritarians, who seek to exert and extend their power by claiming a monopoly over fact and truth. Erasing or rewriting the past allows the tyrant to at once normalize the tyranny of the present and constrict ideas about the future.

It's impossible to talk about the political uses and abuses of memory without mentioning George Orwell. In his classic dystopian novel *1984*, the keeping of history is akin to treason, and one of the main instruments of the police state is enforced forgetting. The novel's protagonist, Winston, struggles to overcome official amnesia: "Beyond the late Fifties, everything faded. . . . Everything melted into mist." The Party's infamous slogan is, "Who controls the past controls the future: who controls the present controls the past."

Real-life authoritarians know this all too well. In today's China, journalist Louisa Lim reports, "history is erased" as part of a larger "degradation of memory." In Vladimir Putin's Russia, the state has imposed "memory laws" that dictate which versions of history are acceptable—and which are illegal. Even in democracies, history is contested terrain. Just think about the attempts to, at best, put a gloss on the history of the January 6 insurrection at the US Capitol or, at worst, to shove the whole episode into what Orwell called the "memory hole." Think about the debates over how to narrate the history of slavery. Between 2022 and 2024, state legislators introduced more than three hundred laws intended to constrict the teaching of American history, especially the teaching of critical race theory. Dozens of those "educational gag orders," as critics call them, have become law. What is that, if not a campaign against memory?

Those who seek to monopolize power have every motive to obliterate historical consciousness. Kundera summed it up: "The struggle of man against power is the struggle of memory against forgetting." When despotism is normalized, resistance is futile.

Unless people remember. While authoritarians seek to use historical erasure as a cudgel, the movements for liberation, democracy, and justice have always recognized that memory can be a shield, a subversive act. One of the most important human rights groups in Vladimir Putin's Russia is called Memorial. In China, "underground historians" have sought to challenge the government's monopoly on history. A leading dissident publication there is the journal *Remembrance*.

In the people's movements of Latin America, the victims of political violence are continually brought from the past to the present. Marchers there might chant, "Óscar Romero está presente" or "Berta Cáceres está presente." (Romero is a slain Salvadoran bishop and Cáceres an assassinated Honduran environmental leader; this very sentence is an act of making them present.) Similarly, when the racial justice uprisings shook America in the summer of 2020, protesters shouted, "Breonna Taylor: Say Her Name" and "George Floyd: Say His Name."

All of this is as true for natural history as it is for human history. Like it or not, nature is political. The fate of the living world is a contested terrain between opposing forces—the exploiters versus the preservationists. And nature is not only caught in the currents of history but—like any persecuted minority—is also at risk of being disappeared from history. What we decide to record in the annals of natural history will shape what we decide to protect. What we allow to slip beyond the pages of natural history will be at greater risk of being lost forever. Forgetting is an accomplice to extinction every bit as much as it's an accessory to humans' oppression of other humans.

Consider today's world from the perspective of the sea star or the firefly and it looks like oppression pure and simple. Most of the world's creatures live in what, if only they could speak to us, they might call a dictatorship of the *Homo sapiens*. Amid such oppression, any contribution one can make to the writing of natural history becomes an act of disobedience to the forces of destruction. It is the citizen naturalist's duty to remember the once-common monarch butterfly. It's the community scientist's responsibility to make a record of shrinking winter.

La vaquita está presente. The passenger pigeon—say its name.

Community science and the writing of natural history are democratic practices, accessible to most everyone. But let's be careful not to overestimate how easy that work can be. I've also seen how challenging it is to transform the act of witnessing into a record that stands the test of time, and I would be remiss not to mention it. Spoiler alert: it doesn't have a happy ending.

The story I'm thinking of has to do with the comparative photography of glaciers over time.

Like a centuries-old tree, glaciers are a door to other timescales. Across a range of cultures—from the Swiss Alps to the Himalayas to the Canadian Yukon—glaciers have long been symbols of permanence. That cultural baseline is now on the move as glaciers around the world melt fast due to the continued combustion of coal, oil, and gas. In Iceland, glaciers have shrunk by some 850 square miles just since the turn of this century; in that time, at least sixty small glaciers have vanished. The United States's Glacier National Park in Montana will eventually need a new name.

These kinds of fast changes might be impossible to miss, but they can still bewilder. Think of the phrase *glacial pace*. It has always meant "slow," painfully slow. But on an overheating planet, glacial pace doesn't really make sense anymore—not when a glacier can disappear in less than a single human generation. When words lose their old-fashioned meanings, it makes it even harder for us to comprehend the change. The Anthropocene has made glacial pace obsolete—proof of how the Great Acceleration is moving so fast that it's leaving language in the dust. Some disappearances will become invisible if only because we'll be left without any language to describe them.

But the universal language of photography is easy to understand. For some 140 years, people have been taking pictures of glaciers from the same viewpoint, decade after decade. Such re-photo projects make for rock-solid evidence of how Earth is being transformed by climate change. It's hard to argue with a picture that's right before your eyes.

One of the longest-running comparative-photography studies of glaciers has taken place on California's Mount Shasta, at the southernmost tip of the Cascade Range. Mount Shasta erupts from the landscape, a solo peak with no other mountains nearby to obscure its height or provide any sense of scale. One can spot the summit from a hundred miles to the south, its snow-white tip soaring above the oak savannah. Comparison to *The Hobbit*'s Lonely Mountain is inescapable; someone has even installed a twenty-foot-long metal dragon sculpture on the plains to the north of the peak.

Shasta is home to the longest glacier in California, which is also the first glacier in the continental United States to have been positively identified by modern scientists. The Whitney Glacier was christened by Clarence King, who led the first US Geological Survey party up the summit and named the formation after Josiah Whitney, a preeminent nineteenth-century geologist. A "colossal cone of a burned-out volcano, springs upward from a plain . . . bearing upon its summit a cap of solid white" is how King described Shasta in *The Atlantic Monthly*. There is a photo dated September 11, 1870, of the Whitney Glacier, its mounds of ice split with giant crevasses.

This combination of physical magnetism and foundational natural history is what drew Phil Rhodes to the mountain in the early 1970s. Tall, lanky, and earnest—and now in his seventies—Phil grew up in the Marin County suburb of Mill Valley, where he developed an early love of wild nature while exploring Corte Madera Creek as a child. He became a member of the Sierra Club when he was fifteen years old and went on to study geology at Humboldt State University (today's Cal Poly Humboldt). "I was very much into mountaineering and backpacking and hiking," he told me. "I fit the picture of the classic Sierra Clubber, you know, back in the old days, with the camping out and climbing mountains and all that stuff." It's an image that holds true today, with his dun-colored cargo pants and beat-up California State Parks baseball cap.

Phil made his first trip to the Whitney Glacier in 1972 and was hooked immediately. "I quickly became aware that the glaciers on Mount Shasta were acting in a very unusual manner," he said. "And that led into a lot of scientific investigation as to why the glaciers were behaving the way they were."

Inspired by a book titled *Exploring Glaciers—With a Camera*, Phil set up his first photo station near the Whitney Glacier in 1974. He then established more Whitney photo stations: W2, W3, W4. He eventually set up additional photo sites to capture some of the mountain's other glaciers—the Bolam, the Holtum, the Wintun—for a total of fourteen locations. He thought of his work as a continuation of the history of amateur scientists who have studied glaciers, most obviously John Muir, people "who are considered to have a scientific expertise, but were not, technically speaking, professionals."

His passion became something of an obsession. From 1983 to 1988, he lived in the village of Mount Shasta and passed the summers as the caretaker of a hut that serves as a gateway for mountaineers attempting the summit. He spent a lot of time searching out old-timers in the area who could tell him more about the mountain's past, people like Harry Watkins, a local lumbermill worker who had taken his own glacier photos in the 1950s. (Today, a glacier on the mountain is named after Watkins.) Season after season, Rhodes kept returning to his photo stations to build up his record of images.

Even after Phil moved back to the San Francisco Bay Area to manage his family's shipping-export business, he sustained the endeavor, faithfully making the four-and-a-half-hour drive northward by himself, climbing the mountain, usually alone, to reach his photo stations. He was driven, in part, by a scientist's commitment to gathering as much data as possible. "What I did was try to make this definable by setting up these photo stations: taking the same photo year after year after year," he said in one of our conversations. "You can look at these photos, and you can see changes."

Rhodes was also motivated by an instinct for historical preservation. "I'm very much into history," he said. "Everything I do is for posterity. To the extent you can retain something, my desire is to retain it." He carries around with him, at all times, several thumb drives with his years of glacial photography backed up. Just in case, he says, his old house burns down—then the record of his life's work won't be lost.

"What I like to do more than anything is to remind people of the way things were, hoping they can learn from that," he said.

"I'm old enough to remember the world fifty years ago and how it has changed—and all the devastation, from the forests in the Sierra Nevada to the glaciers. People will say, 'Are you living in the past?' No—it's really to make the present and the future better."

Eventually life got in the way of the glacier photography—the demands of running a business, the responsibilities of raising a family. In the 2000s, he handed off the project to Hassan Basagic, who at that time was a graduate student in the Portland State Geology Department. And then, sometime around 2015—no one can precisely remember the year—Basagic handed it over to some US Forest Service rangers who were eager for a diversion.

And that's how I ended up in the company of a Forest Service seasonal ranger named Forrest Coots, who did me a big favor by agreeing to take me up the mountain one spring weekend so that I could see a photo station for myself.

A former professional skier and climber who had been sponsored by the gear company Arc'teryx, Forrest grew up in the area, and he knew the mountain better than perhaps anyone else alive. His Mount Shasta baseline starts with his earliest memories, and as we bounced up the furrowed backroads on the mountain's north side, the ecological reminiscences poured out of him. He was born in 1981, and in just over forty years he had witnessed so many changes. The winters had gotten shorter and milder, the precipitation patterns weirder. "I remember, as a kid, you never doubted that winter would come." And now? One recent winter, he said, "We went thirty-six days without any precipitation. Like, dry as a bone." At the same time, the fires had gotten more intense, such as the 2021 Lava Fire that had burned at elevations that never used to see fire. "If you had told me that a forest fire would be at eight thousand feet on Mount Shasta, I'd be like, 'Yeah—*no.*'"

The dirt road petered out at the edge of a sandy wash, not far from a faded sign delineating the boundary of the Mount Shasta Wilderness. We pulled on our boots, strapped on our packs, and started up the mountain. For a mile or so, we followed the faintest outline of a path, and then we were totally off-trail. Nearly all climbers approach the Shasta summit from the south, and the northside was an empty, trackless wilderness. Forrest navigated us upslope through a combination of mountaineer's instinct and memory. We wended up and down gullies and avalanche chutes and through dense stands of pine, many of which were dead or stressed, their needle tips brown and crisp.

He pointed to the southeast. "That's the Bolam Glacier. I doubt it will be here at the end of my lifetime. The Hopland—maybe. The Whitney—maybe. But not the Bolam. It'll be gone."

When Forrest began taking the glacier photos, it was mostly as an escape from his regular responsibilities of rescuing climbers and hikers. He kept at it in part out of a sense of loyalty to Phil. "He is very passionate about it, which inspires us to keep going," Forrest said as we humped upward through scree and muddy snow. "Phil is so passionate that it makes me want to be like, 'Okay, we're going to try to get all the photos.'" Eventually, his Forest Service supervisors made it one of his official duties. "Now, it's part of my performance evaluations: *Did I get it all done?*"

We were now well past the tree line, in a sere landscape of dirt, rock, and ice. We had been climbing for some three hours and were about three thousand feet higher than where we'd left the truck. Forrest guided us to a bench of rock above a tiny ice pond. The Whitney Glacier flowed above us, its ice bright white and digital blue. Beyond that was the flat-topped summit of Shastina, a separate volcanic cone that's known as Shasta's little sibling. "We're here," he said. "This is the spot."

It took me a beat to understand. Forrest pointed to a salmon-colored, medium-sized boulder at the edge of the shelf. Photo station Whitney #1, it turned out, was a completely anonymous, completely ordinary rock situated in the middle of nowhere. It could've been any stone in the world.

This one perfectly average boulder had, for decades, been a vantage point for witnessing and recording the world transforming. But the solid rock was ephemeral. An anchor to understanding the history of one of the most iconic peaks in California was held by little more than the thin thread of lived memory.

After a below-freezing night at our rocky campsite beneath the Whitney Glacier, Forrest and I began our climb up the ice toward the summit of Shastina. I've spent more time with glaciers on the page than I have underfoot, and I marveled at the uncanniness of a glacier traverse: the giddiness of walking on ice that's akin to rock, the strangeness of being on terra firma that isn't firm at all. We hiked past great slabs of multiyear ice tumbled into fortifications—white redoubts and balustrades of blue. Crossing an ice bridge, I paused to look down into the pulsing abyss. The earth here was moving, however slowly, beneath our feet.

During my time with Forrest, I had been able to piece together more of the story of the Shasta glaciers comparative photography project. In many ways, it was an admirable tale of citizen science and bootstrap history. It was humbling to consider the mid-twentieth-century work of Harry Watkins, who did so much amateur study of the mountain's glaciers that the government eventually named one after him. It was inspiring to think about how Phil Rhodes ("the foremost expert on Shasta's glaciers," in Forrest's words) had picked up the effort in the 1970s and then spent so much energy making a record of these vanishing forces of nature. All of that work on his own dime and in his spare time.

This, I imagined, is how memory holds on from one generation to the next—through the often-thankless work of people trying to preserve remembrances worth keeping. Except that sometimes memory doesn't hold. A great many of the Shasta glacier photos, it turned out, had gotten lost. I learned that nearly all the Shasta glacier photos from 2015 to 2019 were missing. It was a loss that, Forrest said, had left Phil "distraught."

And then the project slipped still further. The summer after he guided me up the mountain, Forrest quit his Forest Service job. The comparative photography effort "fell apart," as Phil put it to me. In the fall of 2024, the Forest Service took images of the Hopland, Konwankiton, and Wintun glaciers, but never posted them online. The photo station at the base of the Whitney Glacier reverted back to what it had been for millennia: a plain old rock.

At one point, a partial set of Phil's Mount Shasta photos had been transferred to Andrew Fountain, an eminent glaciologist at Portland State University. But even that was a kind of evidentiary purgatory. Andrew told me that over the years he has collected more than ten thousand glacier photos. Some of them can be viewed on his website, *Glaciers of the American West*. Many of them, however, sit unseen on data servers and hard drives.

And while *Glaciers of the American West* does allow people a chance to glimpse glacier loss via comparative photography, the site is pretty janky. The interface is clunky, the images are hard to navigate, and the site hasn't been updated in years—the result, Andrew said, of a lack of resources, especially now that he's retired. If you go to the site's Glacier RePhoto Project and click on the photo gallery for Mount Shasta, you get a 404 error message. The galleries of Mount Hood and Mount Rainier are also busted. "I'd like to get that site going again," he told me, a weariness in his voice.

If a glacier melts off a mountain, and there's no publicly available visual evidence of it, did it ever exist?

Andrew told me that for much of his career he had avoided such thinking. He approached his research with a "kind of academic remove." Eventually, though, he realized that "in the near future, like maybe 2070 or whatever, glaciers will get so small that nobody will read my work." The realization led him to "a certain depression."

"Surprisingly, not that long ago, only a few years ago, it dawned on me that my work could go the way of the dodo bird," he said. "In a way, frankly, it is encouraging me to work a little more feverishly, because right now what I'm doing is developing an inventory of all the glaciers in the western US, trying to tie down where they all are and how big they are. It might be a map of what times used to be."

I remembered how when Forrest and I were clambering down off the Whitney Glacier, the sound of spilling water was all around us. The spring melt was underway, and countless freshets flowed downhill. The glacier was melting beneath our feet—and we were the only ones there to experience it.

A story remains alive only so long as it's retold. Our histories are a lot like glaciers: they are living only if they stay on the move.

CHAPTER 4

PASS IT ON

Jewell Praying Wolf James was set to go fishing, so if we wanted to talk, I'd have to meet him at the river. The previous few days had been packed with rain and the Nooksack was running fast and strong, the currents aswirl in a chocolate-milk-like flood. It was late autumn, and the cottonwood and willow were bare. Dogwood blared red. Snow geese and tundra swans scratched the sky in tattered *V*s, southward bound.

The Nooksack River forms the eastern boundary of the Lummi Nation, one of the most populous Native tribes in the state of Washington. The reservation is located on a peninsula that elbows into Bellingham Bay, a slip of woods and wetlands fringed with gravel beaches. From Gooseberry Point—a place inhabited for thousands of years—the western horizon is crammed with forested mountaintops, every one of them an island. On that cool fall day, the river was busy with Lummi (pronounced lum-EE) fishermen. They zipped across the water in small aluminum skiffs setting drift nets to catch chum salmon on their spawning run.

I live ten miles from the reservation boundary, and I've often walked the riverbank there. I had heard Lummi folks refer to the place as "Stick's Drag," for the logjams that pile up at the river bend. Jewell had another name: "Outlaw's Drag," a memory from when state authorities hounded Native tribes to try to prevent them from exercising their treaty rights to hunt and fish. When he was a kid in the 1950s, Jewell remembered, there was a village on the riverbank—about twenty families, plus a couple of fish houses that bought and sold the catch. Jewell and the other children were paid a few pennies each to help pull the nets out of the water. "But I decided not to be a fisherman," he told me, "because there were too many fishermen and not enough fish."

Jewell had come to the river for some recreational fishing. The day before, he had filled a five-gallon bucket with ooligans, a small smelt that, deep-fried, make for a crunchy delicacy. Now, Jewell was hoping to repeat his performance. When I arrived, he was standing in the shallows in knee-high rubber boots, plying the water with a metal dip net attached to an eight-foot-long aluminum pole. He had cinched a rope around his waist and tied it off to a riverbank alder; falling into that cold, swift water would be trouble. His bucket held just three finger-sized fish.

Jewell, who also goes by the Lummi name tse-Sealth, was annoyed—less at the paltry catch than at the other ooligan dipnetters working upstream. The elders would not have been pleased. "You see the people dipping ooligans there? Well, this is the spawning grounds of the ooligans. They don't go any further. And my grandfather says we never allowed anybody to drift in between the two bridges. They're dragging their nets over the ooligans' spawning grounds and so there's hardly any ooligans, you see?"

For years Jewell directed the Lummi Sovereignty and Treaty Protection Office, and as he took a break from dip-netting to eat a sandwich of salmon salad on wheat bread he launched into an energetic monologue about Native rights. Jewell briefly went to law school before returning to the reservation to assist the Lummi's sovereignty advocacy, and he's spent time in rooms at the White House and the United Nations. He has the affect of a seasoned litigator, his talk a whirlwind of statutes and citations and court decisions—all of it delivered with the heat of a preacher.

"I'm always an aggressive speaker when it comes to those issues," he said. A leader from another tribal nation once called him "the angry voice behind the movement." "I was aggressive as a child, and rather than turn that aggression inward, I'm going to turn it outward in defense of my people." The passion has kept him youthful: though in his seventies, his hair, worn long in a braid, remains mostly black, touched with bright silver at the temples.

Jewell's uncompromising advocacy for Indigenous sovereignty also comes in the form of storytelling. He is the master artist of the House of Tears carvers, a Lummi collective that specializes in creating massive totem poles that fuse art and activism. Over the decades, Jewell has carved poles to oppose tar sands mining, to try to liberate an orca held captive at an amusement park in Florida, and to commemorate the victims of the 9/11 attacks—a pole that stands in the Congressional Cemetery.

While non-Natives commonly refer to the carved artwork as totem poles, Jewell prefers "story poles." I had first seen one years earlier, when I met Jewell as he was preparing to take a twenty-five-foot-tall story pole carved from a four-hundred-year-old cedar on a cross-country trip to oppose extractive projects like oil and gas pipelines. Painted in turquoise and red, gilded with copper, it was a polyglot of symbols from various cultures: bear, salmon, orca, moon.

"It's a collection of stories from around the world," Jewell told me at that time. "It basically shows that red, black, white, and yellow were all Indigenous. At one time, we were all extremely connected to the earth and taught our children how to care about the environment we live within." Along its journey, crowds of people came to see the pole and lay their hands upon it, the carving acting as a kind of megaphone broadcasting messages of resistance to environmental destruction.

Now, on the banks of the Nooksack, Jewell had another story to share. He lit up a cigarette and I noticed that the fingers of his left hand were dotted with the blue ink of a homemade tattoo: L-O-V-E.

"I used to be head of [Lummi Nation] fisheries. From '79 to '84. When I was fisheries director, the fishermen were upset because I had to close down the spring Chinook fishing, because there was hardly any spring Chinook. I was only twenty-six at the time, but I had enough sense to know that we can't keep destroying the population without consequences."

The Lummi fishermen were livid. "They felt like I was an asshole, that I didn't know what the hell I was doing."

There was a big meeting. Jewell got up and, in front of everyone, told the story of Salmon Woman. The legend is a foundation of Lummi culture, and Jewell's capsule version of the story goes something like this:

> The People were starving and so Raven, which is the leader, found Salmon Woman when he was lost in a fog. And Salmon Woman, knowing the People were starving, gave her salmon children to Raven's people. And she sang a song to lift the fog and have her children follow her to the rivers. She brought the salmon to the rivers. Once she did, the People were able to survive. All Salmon Woman wanted was for us to respect her children. So we're supposed to return their bones to the water and as a consequence their spirit would return to the house of the Salmon Chief, and then they would come back next year.

The fishermen weren't having it. "We're hungry!" they yelled. "We've got to feed our families, you bastard, and here you are standing there and telling us children's stories!"

Jewell thought he'd have a minute to interpret the tale's meaning: the necessity of protecting the salmon, protecting the river, protecting the spawning beds. But the crowd was getting rowdy. It was becoming, he said, a "let's hang him type of moment."

Then an elder, the late George Cagey, got up to speak. He told the room, "You're right! We have the right to take those spring Chinook. Let's go to the river and take those spring Chinook. And when we take everyone last one, let's go to the spawning beds and stomp on the eggs." Everyone in the crowd hung their heads, went home, and accepted the fisheries' closure.

"He made them realize what the story was all about," Jewell said. "It's our duty to protect the salmon and the habitat."

The Pacific Northwest salmon are the stuff of legend. They are also a textbook-worthy example of environmental generational amnesia. To read the salmon literature is to enter a library of shifting baselines, to see, once more, how easily the meaning of scarce can shrink even as the definition of abundant shrivels.

For thousands upon thousands of years, they came by the tens of millions from the sea to the rivers: Chinook, coho, sockeye, pinks, chum. Every creek across greater Cascadia had its own genetically distinct population of salmonoid, the fish home to their one stream and no other. It was a vast wheel of life churning between salt water and soil, the nutrients of the ocean cycled upstream to feed the forest and all its inhabitants. After spending their adult years in the Pacific, they navigated back to their natal waters by using the planet's magnetic field like a compass as well as by scent memory. It was Proust-like—the fish returning to their childhood home by the lure of smell.

When the first Europeans arrived at the Salish Sea—the inland salt waters that stretch from present-day Olympia, Washington, to Campbell River, British Columbia, and which includes the Puget Sound, the Strait of Juan de Fuca, and the Strait of Georgia—they were gobsmacked by the fairy-tale beauty. Glacier-armored peaks towering above primordial forests, meadows dressing the riverbanks. "The country before us exhibited everything that bounteous nature could be expected to draw into one point of view," British navigator George Vancouver wrote upon his arrival. A century later, John Muir called the sea "the Mediterranean of the North American continent."

Most of all, the newcomers marveled at the fish. The once-great salmon runs of Europe were nothing more than relics; the salmon fisheries of North America's Atlantic Coast would soon fall, too, smashed under the heel of nascent industrialism. But in the Pacific Northwest plentitude remained. When Lewis and Clark arrived on the Columbia River in 1805, they wouldn't even attempt to estimate the salmon: "The number is incredible to say." In the 1840s, a pioneer in what today is Oregon wrote, "I never saw so many fish collected together before." The settlers were astounded by the fishes' size as much as by their sheer numbers. Hundred-pound Chinook were not uncommon; "June hogs," they were called.

It didn't take long for amazement to curdle into avarice.

Within just a few generations, the settlers transformed the sea's natural wealth into cold, hard cash in a spasm of destruction.

The first white settlers had shown little interest in salmon (their attention was focused on otter furs and land grabs) but with the invention of fish canning their hunger fell on the sea. The largest salmon-processing facility in the world sprang up in the Fairhaven district of Bellingham—not far from where I live today—and by the turn of the last century it was annually packing five million one-pound cans of salmon. As lumberjacks converted ancient groves of cedar and fir to board feet of lumber, the hillsides lost their footing and suffocated the spawning streams with silt and mud. At the river mouths, pulp and paper mills asphyxiated the coastal waters with sawdust.

The truly fatal blow arrived in in the mid-twentieth century in the form of hydroelectric dams. Concrete curtains descended across Cascadia. On the Columbia River, the Grand Coulee Dam was completed in 1942 without any fish ladders, annihilating a thousand miles of spawning habitat. After dams went up on the Yakima River, the number of annual sockeye spawners collapsed from some six hundred thousand to scant thousands. The great wheel of life between headwaters and ocean was jammed shut.

For more than a century, people in the Pacific Northwest have been worrying about the unmistakable decline of salmon. Then each new generation redefines "decline" downward. There's always something worse than worst.

The Pacific salmon population collapsed by 50 percent between 1849 to 1980, around the time Bruce Brown wrote a book called *Mountain in the Clouds* in which he bemoaned that the likeliest place to find a wild salmon in Washington was in "the state archives." Since then, twenty-eight species of salmon and steelhead have been placed on the endangered species list. In his 1990 paean to Cascadia, *The Good Rain*, Timothy Egan laments how the sockeye runs on the Fraser River had fallen from a historic high 40 million to a mere 2 million. Thirty years later, the 2020 sockeye run on the Fraser was a measly 280,000 fish. History repeated itself–and then it compounded.

The ever-present state of emergency seems to have anesthetized our discernment. How else to explain a celebratory *Seattle Times* article about the return of five thousand Chinook to the city's Lake Washington? "A splendid abundance," the newspaper called it, even though the figure was a small fraction of the historic high. Some years ago, the Associated Press trumpeted that the fall Chinook on the Snake River had made a "comeback." The run was about seventy-five thousand; the run in the 1930s was more than six times larger.

Recall that the concept of shifting baseline syndrome sprang from fisheries science. The latest peer-reviewed evidence comes from the salmon-spawning rivers of Idaho, where many runs are classified as endangered or threatened. The Chinook population in Idaho rivers is about 3 percent of what it was in the 1950s, and the government fisheries managers have set 10 percent of the 1950s baseline as their recovery goal. But, according to one group of government scientists, in the 1950s the fish numbers were just one-third of what they had been in the 1880s. "Even modest increases in salmon returns are incorrectly interpreted as 'record runs' despite being far below historical abundances," the researchers warn.

It all comes back to memory. In *Salmon*, Mark Kurlansky spends nearly four hundred pages detailing the long train of abuses the fish have suffered and then, at the very end of the book, indicts shifting baseline syndrome as a main culprit preventing real restoration. "If the goal is to get a river back to its disastrous level of 1950 or even 1850, the salmon will not be restored. The problem is that no one remembers the abundance of salmon before the Industrial Revolution and so they cannot aspire to it."

When a fraction of a fraction is considered success, the definition of environmental restoration has been rendered into garbage. It's as if you awoke from horrible car crash to be consoled by the doctor that you'll only be paralyzed from the neck down: it might be survival, but you'd never call it recovery.

While it's true that many people have forgotten the salmon's former abundance, some have managed to remember. Steve Solomon is one of them—and he's not alone.

Steve is a Lummi fisherman who spends as much time as he can on his wooden, red-painted skiff setting nets on the Nooksack. In recent years, he's passed most of his time on the beach due to low fish numbers and fisheries closures. "We went from fishing spring, summer, and fall to just fishing, in essence, a fall season," he told me one autumn evening. "Ninety days, and we're sitting on the beach most of it."

Steve—seventy-one, with a face and hands that telegraph decades of work on the water—has a lifetime of fish tales. When he was a kid, the herring fishery was still booming and the salmon fishery, too. "There was lots of fish back then," he said. "Bellingham Bay, at the peak of the Chinook season, it would be so many lights and boats out there that it looked like a city. You could literally jump from boat to boat." He can remember the relatively good years of the late seventies, when the Lummi had one of the largest Native fishing fleets in the United States. "In 1978, our fishermen brought into this tribe $17 million from the sockeye."

He holds, too, a storehouse of stories his grandfather left to him. His grandfather told him how, in 1905, he traveled to his ancestral reefnet fishing sites in the islands—the reefnet being a unique fishing technology invented by the Lummi—to find all of them blocked by settlers' fish traps. "There was a trap at every site he owned," Steve told me. "And he says, 'What am I going to do for my people and my family?' He says, 'I have no means to support them.'"

The canneries offered his grandfather a job. "That job entailed throwing all the fish they can't use into Georgia Strait, in the cover of darkness. By the scow loads. And my grandpa would cry telling me this. He says, 'Sonny, all they took was the bellies. All they took was the bellies, and I had to throw them overboard. Throw them overboard by the thousands.' And he told the owner, he says, 'You people can't continue to take and take and take. You're going to kill it.'"

Steve has stories older still. His great-grandfather—with whom he shares his Indian name, Tla kalin Ces-xen—was born at the large Lummi village located at what today is called Whatcom Creek, originally *Xwot'qom*, meaning "noisy waters." Before white settlement—before it became the site of a Vietnamese restaurant, an antique store, a couple of cafés, and a dive bar—three longhouses stood there. Steve's great-grandfather grew up in one of them. To picture it, to knead inherited memory into daydreams of the past, conjures the good times before the aftermath. "How plentiful the fish were at every creek, at every little side stream, every good-sized creek that could support all these salmon."

The Lummi call themselves Lhaq'temish—the People of the Sea. Their traditional territory was a water world: islands beyond islands, green humps of forest sprinkled among the waves. Before European contact, the sea provided nearly everything. They ate twenty different kinds of fish, consumed ten species of oysters, clams, and mussels. *When the tide is out, the table is set*, an old proverb goes. The sloshing rhythm of ebb and flow formed the backbeat of the universe.

"We harvested for three months and celebrated for nine," Jewell told me on the riverbank. "Nine months of spiritual, cultural, ceremonial, ritual sharing." The Lummi and other Coast Salish nations were so wealthy that their most celebrated ceremony was the potlatch—gatherings in which the host gives away food and prized possessions. In the Lummi dialect, there is no word for famine.

"You got to keep that story alive, because it once was thriving, it once was happiness," Steve told me. "Sharing stories—it just needs to happen because the fish are in trouble. What are you going to do to help? Our brother, the fish, is crying for help."

I have heard similar talk across Salmon Nation—the string of Indigenous societies that stretches from Alaska to Northern California and whose cultures are linked in familial relationship to the salmon. The epic runs may be long gone, but they're not forgotten.

Shannon Wheeler, chairman of the Nez Perce, once told me that "it's easy to keep the salmon memory alive, because it's an obligation. The salmon gives its life to us, so we're supposed to give back." JoDe Goudy, a former chair of the Yakama Nation, one time said to me, "We are talking about a right and respectful and reciprocal relationship between people and the river and the salmon that has been maintained for thousands of years."

From the moment they signed treaties with the US government, the tribes of Salmon Nation have fought for the salmon. Most often, those fights have taken place in the courts. In case after case, federal judges have upheld Native treaty rights to fish as they always have. When the courts have been too sluggish—or the authorities too recalcitrant or downright racist to enforce court orders—Native peoples have engaged in direct action to protect salmon and sovereignty.

The most famous episode was the Fish Wars of the 1960s and 1970s. Led by the charismatic Nisqually fisherman Billy Frank Jr., tribal members across Washington were repeatedly arrested as they deliberately violated state fishing regulations to assert their treaty rights. Things got intense. Authorities clubbed Indigenous activists and launched tear gas into a riverside protest encampment outside Tacoma. Native protesters set fire to a railroad trestle. Shots were fired by both sides.

The tribes won the war. In a landmark 1974 decision, federal judge George Boldt ruled that tribal members should be guaranteed the opportunity to take 50 percent of Washington's salmon catch. The Boldt decision was based on a plain reading of the 1850s treaties, which secured the tribes' "right of taking fish, at all usual and accustomed grounds and stations . . . in common with all citizens of the Territory." The sea, it turned out, couldn't be enclosed like a terrestrial reservation. *All* the salt water constituted Natives' "usual and accustomed places."

But as Steve Solomon likes to say, "A treaty right without salmon is not a treaty right at all." Half of zero is zilch.

With salmon numbers still so achingly low, the tribes of Salmon Nation have continued their advocacy on behalf of a species they consider to be kin. On the Olympic Peninsula, the Lower Elwha Klallam Tribe forced the demolition of two dams; coho fishing for tribal use resumed in 2023 for the first time in more than a century. On California's Klamath River, the Yurok and Klamath nations spearheaded a decade-long campaign to pull down four dams; within weeks of the dams' 2024 removal, Chinook were spotted spawning. Across Salmon Nation, dozens of tribally managed hatcheries are providing the fish with a lifeline until wild salmon stocks can be restored. Whatever salmon remain in the rivers of Cascadia are there thanks to Indigenous leadership.

Memory forms the foundation of this generation-spanning defiance. Any sense of injustice rests on remembering—a recollection of treaties broken, promises unfulfilled, common wealth sacrificed for individual profit. "The Yakama old-timers, they can still hear Celilo Falls," JoDe Goudy told me once, referring to the Yakama fishing site at the falls that was continuously inhabited for eleven thousand years before it was drowned by the Dalles Dam. "They can hear the echoes."

Those echoes—those memories—remain alive today because they've been passed on. And most often they've been passed on via the old ways, in the form of spoken stories. For the first peoples of this continent, oral history is at once the anchor of identity and the sail that powers their political struggle. Oral history is a cultural commons that can never be owned—memory held in mutual trust. "One thing about Indian Country is that most of the elders have a teaching: *These are borrowed words*," Jewell told me.

The Native histories and perspectives here are also borrowed words. They were lent to me in the expectation that I would share them. Having received the story, I have a responsibility to pass it on.

Stories forestall forgetting. To remember something clearly, cast it into narrative. One of the best ways of sharing stories is one of the oldest: word of mouth. When it comes to preserving ecological memories hitched to time and place, oral storytelling is so valuable because it's so easy; storytelling is perhaps the most universal of human arts. Compared to oral histories, clay tablets stamped with cuneiform are mere infants.

In preliterate cultures, all information—from the routes between locations to the uses of plants and animals to the moral codes that undergird social well-being—had to be held in the mind. Nowadays, in contrast, people know how to record everything and seem to remember nothing. Also, the vicissitudes of fate can be unkind to the written record: libraries burn, scrolls get lost forever, websites go dark, artificial intelligence hallucinates. Oral histories, by comparison, are strikingly durable. They can hold strong over thousands of years. And what oral histories might lack in precision they more than make up for in longevity. Oral histories provide ecological baselines that stretch to time immemorial.

Consider "geomyths." Geomyths are folklore and other legends that provide explanations for natural features or geological events that happened long ago. A good example is the flood myth of Noah's ark or the lost realm of Atlantis. Those tales may have been seeded by a vast deluge that occurred more than eight thousand years ago when the Mediterranean spilled into what is now the Black Sea. In Australia, Aboriginal "songlines"—oral histories that connect landscape, plants and animals, and people—tell of "the time before the sea" and refer to dozens of specific locations that have been submerged. Modern science confirms the ancient lore: the first humans arrived in Australia some 65,000 years ago, when ocean levels were much lower and the shoreline dramatically different.

Indigenous stories are replete with geomyths—not at all surprising when you consider that Indians were the real pioneers, the original discoverers of a truly new world. The Zuni people tell of a long-lost epoch before humans, a time when all the world was ocean filled with sea monsters—until a massive conflagration dried out the sea. And, indeed, in the Cretaceous Period much of the American Southwest was a great inland sea. When the first humans arrived, they would've encountered marine dinosaur fossils. The Zuni tale is paleontology in the form of creation myth.

Oral histories from the Klamath Tribes of southern Oregon refer to the volcanic eruption of Mount Mazama that formed the caldera of Crater Lake. The explosion happened 7,700 years ago. That means that eyewitness accounts of the eruption have been handed from elder to youth over some 250 human generations.

The Lummi story of Salmon Woman could also be a geomyth. Until about 15,000 years ago, the remnant glaciers of the Ice Age separated the Pacific Ocean from the rivers that today feed the Salish Sea; it wasn't until the ice melted that salmon colonized the interior. And that was just about the time that the first humans arrived in the region, according to the latest archaeological findings that suggest some Paleolithic people came to North America via canoes on a "Kelp Highway."

"So, how old is the story?" Jewell said. "Well, old enough to know that we were here before the salmon arrived. And once they did, we were able to survive."

Geomyths are memories of Earth's prehistoric past compressed into legend by the weight of time.

Maybe this seems like wizardry. Anyone who has ever played a parlor game of Telephone knows how easily oral transmission goes awry. How is it possible for oral history to convey facts over the arc of millennia?

In his book *Memory in Oral Traditions*, David C. Rubin, a cognitive psychologist, provides a detailed explanation for oral history's impressive stability. Using Balkan poetic epics as his primary examples, Rubin explains that one major reason for the durability of oral transmission is that it works more like a web than a chain. The tales are performed publicly, which means that over time there are many speakers and many listeners, instead of the individual-to-individual transmission of ordinary gossip or a game of Telephone. Like any cherished reminiscence, oral traditions are long-rehearsed and retrieved again and again. Like a flashbulb memory, the plots are usually simple and infused with vivid imagery—*she sang a song to lift the fog*. Alliteration, assonance, and rhyme all help. So does adaptability. The singer of the epic poem and the folk tale raconteur often swap lines from one telling to another, and those tellings are often of widely varying lengths. There were many versions of *The Iliad*; only one was set to writing.

This might seem too obvious to state—a bit of tautology—but oral history is durable precisely because it's spoken aloud to someone. Truly, the medium is the message. Whereas reading is (almost always) a private endeavor conducted in silence, any oral storytelling is a social experience.

The spoken word draws speaker and listener into relationship. The experience is rich with associations. It may be that one of the reasons why oral history sticks so tightly to the mind is because it taps into both semantic memory and episodic memory. There's the *what*-memory of the story itself and, layered on top of that, the *who-when-where* of the storytelling event. Memory is more process than thing, more verb than noun—just like the act of storytelling.

Above all, oral histories are stable because they are infused with meaning. "A listener does not hear a song in isolation," Rubin writes, "but in relation to past events"—especially other songs and stories from within that tradition. A story's meaning comes from its culture. Tell me what a society remembers by way of its stories, and I can tell you what it values.

Of course, a good story can also be bullshit. Some literary critics complain that the tidy organization of stories is a mirage. Real life is a chronicle of chaos more than it is a neat arc with beginning-middle-end. Many writers worry, with good reason, that stories are an inadequate response to capitalism's appetite for destruction. Narratives are no match for bulldozers.

Indigenous storytelling represents a muscular rebuttal to such anxieties. For Native communities that have passed through the fires of genocide, the continuity of stories has been nothing less than a survival technique. The strength of Native storytelling is at the heart of what Anishinaabe intellectual Gerald Vizenor calls *survivance*—active resistance that transcends mere survival. "Native stories are the source of survivance," Vizenor has written. "Survivance is an active sense of presence, the continuance of Native stories." Epidemics took countless Indigenous lives, colonizers took Indigenous lands, and government officials took Indigenous children and shipped them off to abusive and sometimes murderous boarding schools—but no one could take their stories.

Jewell James, of course, knows this. Before European contact, high-status people in Lummi society distinguished themselves by knowing the ancient songs and knowing their family histories. Then the colonizers tried to stop such cultural transmission. "It was illegal for us to practice our song, dance, ceremony, and to transfer knowledge—sacred knowledge, Indigenous ways of knowing, traditional knowledge—to the next generation," Jewell said.

Like Native nations across the continent, the Lummi did so anyway. Jewell told me how, during his grandparents' time, the only way the Lummi could practice their ceremonies was in secret, at a little house way back in the woods where they had to post a guard who would raise an alarm in case the local Indian agent was coming. During the most brutal years of colonization, the singing of the old songs was nothing less than obligation. "When you become somebody in your family, you have a duty to transmit knowledge."

Over the course of time—under the weight of time immemorial—that responsibility became central to the culture's very being. In the preface to a collection of Coast Salish oral histories titled *Jesintel* (which means "to learn and grow together"), a group of Lummi authors write, "Gathering and sharing stories is a Coast Salish lifeway." Stories told by the elders, the authors declare, exists outside "the linear and colonial understandings of time."

This is story as circle, story as another mapping of meaning, as a way of resisting chronos's one-way sprint toward the always receding future. "It is," the *Jesintel* authors write, "the understanding that animals, plants, material objects, rivers, mountains, forests, and essentially all aspects of the world, and the whole universe, (be)come together through story." That story rests, elementally, on memory: "The ability to retell, to remember."

At least half the work of oral history is listening; every conversation is a partnership between speaker and audience. If you're a baby boomer or a Gen X elder-in-training, your job is to share your wealth of memories—potlatch in the form of narrative, as it were. If you're part of the millennial or Gen Z generations, your task is to open your ears. To transmit memories of the planet's past requires closing the circle between old and young.

I witnessed such a closure when I met Free and Raven Borsey, Lummi twins in their late twenties who have dedicated themselves to carrying on their culture's traditions. The Borsey boys had a tough childhood. When they were two years old, their mother dropped them off on their aunt's doorstep. They were "really feral," Raven recalled, and the family they were raised in was "very strict in culture, very strict in teachings, very strict on us." When the other kids got to go run and play, Free and Raven were expected to join the grown-ups in the smokehouse to hear the stories and songs—just sit there and listen, the winter gatherings sometimes stretching for hours.

The twins showed youth's usual impatience with the old. "When I was a kid, I'll admit, I was like, 'Oh, this is just Native stuff,'" Free told me. In hindsight, he credits the strict upbringing with helping "to break a cycle of trauma and destruction."

A life-changing experience came when the twins were thirteen and they participated in a traditional canoe journey from Washington's San Juan Islands to Bella Bella, British Columbia. The paddling trip covered 650 miles of the Salish Sea and took thirty-eight days. At times, they were skippering the canoe through fourteen-foot waves. Camped out on beaches, they ate traditional meals like herring eggs served on hemlock boughs and stayed up until midnight learning the protocols of Coast Salish dancing and singing. It was like they were standing in their ancestors' dreams.

Today, they work at a nonprofit group called Children of the Setting Sun that seeks to bridge Western knowledge and Indigenous wisdom; "where science meets spirit" is the organization's unofficial tagline. Much of that work happens through storytelling, including documentary films and podcasts and a scripted TV program that's in the works. Free's title is Environmental Steward; Raven is a Research Fellow at the organization's think tank. They are what you could call culture couriers, working to build connections across generations and between Native and non-Native cultures. Their work, they insist, isn't political activism—it's "a way of life."

Their lifeway of advocacy revolves around sharing the stories that have been shared with them—stories of the orca and the salmon, and "the inherent responsibilities," as both Free and Raven repeatedly refer to them, that comes with the relationships between humans and other beings. "Everything begins with story," Free said. "Our relationship with salmon begins with the story."

"Stories are literally the most powerful weapon on Earth, and that's been true since time immemorial," he continued. "You know, my ancestors struggled hard to make sure that these stories persisted. Who would I be to then give up? So, yeah—that's the weight of storytelling."

To do it right, Raven insisted, "you have to intentionally pass it on." In twin fashion, he then echoed his brother. "Chief Tsi'li'xw [the late Bill James, hereditary head of the Lummi] has a quote. He says: 'It never was truly yours if you don't share it.'"

The Borsey twins give me hope that anyone can find twenty-first-century ways to pass on ecological memories. Raven calls himself a "boundary spanner," and I think we all have the potential to be boundary spanners, each in our fashion. After all, Native communities may be exceptional in their practice of oral history, but they aren't unique.

As a species, we are hardwired for sharing stories. Neuroscientist Charan Ranganath makes this point at the very end of *Why We Remember* as he conjectures that storytelling in the winter of our lives is an example of memory doing exactly what it evolved to do. "The persistence of semantic memory in the face of episodic memory declines may be a clue about the functions of memory in old age, when life is not so much about new learning . . . but about sharing what we've learned." On the final pages of *Searching for Memory*, Daniel Schacter declares that the best "redress [for] cultural amnesia" is for "elderly adults to integrate the past with the present." Conversations between old and young represent "the intergenerational extension of remembering."

What might this look like in terms of combating environmental amnesia? Perhaps the easiest way of transmitting environmental stories is as straightforward as talking about the weather—an old-fashioned pastime that the climate change era has transformed from pleasantry to political act. According to one study, only 35 percent of Americans talk about climate change even "occasionally." One reason people bite their tongues is because they worry that others don't share their views—even though, in fact, a majority of Americans accept the basic science of global warming. And so silence becomes another negative feedback loop of climate change. Climate stories in the form of weather small talk can help break the self-defeating cycle. Simply discussing climate change with friends, family, and acquaintances can make climate change more of a political priority. "Listen first, and then talk," one researcher counsels.

That ethos of listening is at the heart of the oral-history organization StoryCorps, which, since its founding in 2003, has recorded seven hundred thousand conversations of Americans talking about their lives. The recordings, which are kept for posterity by the Library of Congress, is the largest single collection of human voices ever gathered. This audio encyclopedia of twenty-first-century American life has, at times, touched on climate change—now an inescapable part of lived experience. Search the library for conversations on "climate change" or "global warming" and the results include more than two thousand dialogues about how to prepare for climate emergencies, debates about climate science, and discussions on the best ways to psychologically cope with the impacts of the climate crisis. Such audio snippets of ordinary life will, eventually, become an environmental time capsule.

Once again, attention is everything. To best transmit oral histories of climate change and other environmental shifts demands focus—intentionality, to borrow Raven's word. Professors at University of Puerto Rico have launched a formal program to capture Puerto Ricans' memories of climate-change events like the disastrous Hurricane Maria. The University of Kentucky has also begun an oral history project centered on global warming; researchers there say that "oral histories hold the key to recording environmental change." The folks at NOAA have organized an oral history archive in which their staff share their experiences with a changing environment. Children of the Setting Sun hosts something similar at Salmon People gatherings and conferences. The organization sets up a storytelling booth in which people can share their autobiographies and have them video recorded. Like streamside restoration projects that revive salmon spawning grounds, stories can be acts of environmental stewardship, too.

Another grassroots effort at capturing environmental histories is the Climate Stories Project, which offers people a platform to share their climate change reality through music and personal narrative. "The most successful stories," reads the project's open invitation to the public, "weave personal experiences into the larger issue of our changing climate." The project has gathered audio and video from people worldwide who offer their testimonies—who *bear witness*—about how climate change has impacted them. The short recordings, which are available on YouTube and SoundCloud, form an intimate, sometimes wrenching, record of the ways in which global warming is upending people's lives.

In one recorded story, a woman named Sara recalls her childhood growing up on the South Shore of Long Island, where she spent her time swimming and boating and gathering shellfish: "Clams were so plentiful then. I would bring them home and my mother would make baked clams for the family. But today when I visit the same beach, there are no clams. It saddens me to know that new generations of children are missing out." Betsy in Gainesville, Florida, shares her family's story of surviving Hurricane Helene and how an eighty-foot-tall red oak smashed into her home, just barely missing her sleeping four-month-old baby and sending her husband to the hospital with a sliced scalp. An audio recording of Joseph in the Adirondacks offers a recollection of what it was like growing up in the 1950s, when the winters were colder, the ice was plentiful, and backyard hockey a beloved tradition: "Ice is what I remember most about my youth. . . . It was just a way of life I thought would never end." When Joseph returned to the Adirondacks in the nineties after a career in the Air Force, "I noticed that that way of life was gone. If you're an outdoors person you don't need anyone to tell you that the climate is warming in the north country."

To create memory baselines that can withstand the planetary-scale disruptions of the Great Acceleration requires hurdling over the limits of living memory. One of the best ways of doing that is to write a letter to the future—a hybrid approach that combines the spirit of oral history with the form of a writing-based culture. Such letters make for a powerful method of memory transmission. When researchers evaluated eleven different climate-action strategies—actions that had involved nearly sixty thousand people in sixty-three countries—to see which were most effective, they discovered that the most impactful tactic was asking people to write a letter to their descendants in 2055 in which they would share "how you want to be remembered by them and future generations as someone who did their best to ensure a safe, flourishing world."

There's now an advocacy organization—DearTomorrow—that provides people with prompts to write their own letters to the future. Thousands of people have done so, and their letters, much like audio and video recordings, form a chronicle of this bewildering moment in time, this hinge between the climate we've known and the unraveling climate we're bequeathing. In a letter to her daughter, Abby in Montana recalls "the worst wildfire season in Western Montana in over a generation" when her daughter was in utero and goes on to describe "looking out the window in the middle of January to a nearly snow-less valley here in Missoula . . . the winter that hasn't been." Then, in a burst of maternal protectiveness, Abby makes a pledge of resistance: "I refuse to accept this. It won't be easy, but I promise to reject the forces of apathy, isolation, and defeat." This is memory made into a seedbed for political action.

The power of such letters derives, I believe, from people's instincts for inter-generational responsibility, the obligation to leave a place at least as well as you found it. I think something else is at work—a desire to place our individual lives on the spectrum of history, which for most of us means the spectrum of family. In Indian Country, no personal introduction is complete without a recitation of the speaker's ancestry: parents, clan, tribe, place. But the passion for genealogy is universal. Some 70 percent of Americans say it's important to know one's family history, a statistic proven by the cottage industry of family tree software, DNA-tracing companies, and services like Storyworth that help people create memory books of their lives to hand down to descendants. Genealogy is story as family map, a way of making sense of where we fit in the world.

When it comes to passing on oral history, the real lesson to be learned from Native societies is less about form (for storytelling is common to all cultures) or passion (since everyone is eager to share), and more about subject and content—the way in which culture and environment bleed and blend together. "The stories of place are really stories of family," Darrell Hillaire, a longtime Lummi leader and founder of Children of the Setting Sun, told me. That idea of family is an expansive one—the web of familial relationships includes orca, eagle, salmon, bear. "So the place isn't just, you know, *name of street*. It represents something else, because we gathered there, we hunted or fished there, we lived there. And we received those gifts from the other things that live there, whether it be the animals or the water or the wind. That's how we identify with the place. How *it* identified *us*."

This insight about the human kinship with other living beings is available to anyone. The more you inhabit the first antidote to environmental amnesia—Go Outside—the more natural it will feel to graft ecology onto genealogy. No culture has a monopoly on welding lineage to landscape. All sorts of people have found all kinds of ways to make natural history part of family history. Think of farmers, ranchers, hunters, anglers, fisherfolk, sailors, mountain climbers, or modest suburban homesteaders like me—"practical naturalists," in the words of Ralph Waldo Emerson. Ordinary folks at home in the outdoors.

When "grandfather" and "fishing spot" are synonymous—when "the cousins' visit" becomes metonym for "berry picking"—family and landscape and story have fused together. Such fusion etches memory onto the environment, inscribes the earth with meaning to the point of myth. It's a way of making the living world storyful, of making it memorable.

The drums announced that the ceremony was beginning, and that the salmon would soon be arriving. The sound was thunderous, a chest-thumping bass throb bouncing off the walls of the Lummi Nation School's gymnasium. *Boom-boom-boom-boom*, like the smash of the surf against a rocky beach. Singing rose around the drums like a sea swell, dozens of voices chanting in the Lummi dialect the Lhaq'temish national anthem, "The Flood Song."

It was mid-May, and my daughter and I were attending the Lummi First Salmon Ceremony. The ritual—which takes place in various forms across Salmon Nation—has been practiced since time out of mind as a way of giving thanks for the salmon's annual return. In the early to middle twentieth century, the ceremony nearly died out and was mostly practiced at home by individual families. The Lummi brought back their public ceremony about thirty years ago, first as just a small event in the community center, and then, over the years, growing into a huge celebration.

This was one of the biggest yet. The basketball court was packed with sixteen long banquet tables for elders and invited dignitaries. The bleachers were filled with hundreds of people, including schoolkids of all ages. Outside, a crew of guys including Steve Solomon was fire-roasting spring Chinook caught by Lummi tribal members, the wood smoke blossoming into a clear sky. There would be food for a thousand: salmon, of course, plus crab in the shell, steamed mussels, clam chowder, fry bread, rice, potato salad, cakes and cookies. It was abundance to match a potlatch—Thanksgiving through an Indigenous looking glass.

A group of high schoolers had the honor of emceeing the

event. When the opening song and procession was over, the high schoolers invited the tribal chairman to say some words of greeting. Then they passed the mic to three middle school students—a boy and two girls, the girls wearing bright yellow blankets—to share the story of Salmon Woman.

The boy began, "A looooooooooonnnnnnnggg time ago," and the whole crowd chuckled. While the girls kept it straight—"I am the Salmon Woman. These are my children. Take as many as you need."—the boy continued playing for laughs. There were more "yucks" and "likes" and TikTok era affectations ("fiiiisssh!") than in more traditional versions of the story. It seemed to me a perfect tweenager combination of awkward, earnest, and goofy.

After the ceremony, as the banquet was getting underway, I caught up with Jewell James and asked him what he'd thought of the morning. It was good, he said. But as for the telling of Salmon Woman, "I couldn't figure out what they were saying."

Perhaps even prayer and ceremony are susceptible to shifting baseline syndrome, I thought. But my nine-year-old had absolutely no problem following the story. To my surprise, she had already heard the tale at school, where it was taught as part of the state's "Since Time Immemorial" curriculum about Washington's Indigenous cultures. She said she liked the kids' version better. And the moral of the story had come through clearly enough. "Don't take things for granted," was her take. "It teaches you to appreciate."

The most durable legends are protean. Each generation

refashions the story to fit the purpose of its time, but the core meaning stays fixed in place, especially if it's held fast by the gravity of ritual. Memory is a performance–scene enacted again and again in the mind's eye–and in that sense it's not all that different from ceremony, which is one of the most potent ways of creating collective memories. Song and ceremony are another form of embodied muscle memory; ritual is a cultural mnemonic device. When they sliced up the honorary first salmon and passed around pieces to everyone in the gym–tiny bits of salmon in small white paper cups, to be eaten in a single bite–it reminded me, more than anything, of taking communion as a kid. It was like a sacrament.

I think everyone could benefit from ceremonies of remembrance like that. If the worst-case scenarios are right–if we are truly on course for an apocalyptic age of fire and flood–then Indigenous cultures' commitment to story and memory is one of the tools we'll need on the collective life raft. After all, Native societies are *already* postapocalyptic. Their survivance today is due, in part, to the wisdom that sometimes you have to go back to go forward.

The tale of Salmon Woman isn't merely a saga from the mythic past. It's a parable invested with moral teaching, an instruction manual for how to live in a good way. It's a story, anchored in memory, that charts a path to the future.

CONCLUSION

MEMORIES OF THE FUTURE

For hundreds of thousands of years, during those long centuries when the New World was still old, bald eagles lived nearly everywhere in North America, their one and only natural home. Eagles flew over the tidal estuaries of the Atlantic Coast and the rugged seashore of the Pacific. They built their nests in the folds of western basin and range, on the prairie, throughout the north woods. There were likely a half million of them, an eyrie perched every few miles along fish-rich rivers like the Schuylkill, the Hudson, the Columbia, the Colorado. Up to fourteen pounds in weight and with a wingspan that can stretch eight feet from tip to tip, the bald eagle was sovereign of the American sky, the apex predator of the air.

The First Nations perceived a great power in the bird. Many

Native cultures considered bald eagle feathers as sacred and used them in ceremonies or for ritual healing practices. For the Lenape people of the Delaware River region, it was considered a sign of courage and good luck to wear an eagle feather that had been plucked from a living bird. In the Southwest, the Pueblo and Zuni kept domesticated eagles in large aviaries. Across diverse cultures, Indigenous lore esteemed the eagle as a chief, the bird of heaven. "In an eagle there is all the wisdom in the world," the twentieth-century Lakota spiritual leader Lame Deer (aka John Fire) once said. "Eagle" was very likely the most common animal clan name continent-wide.

When the newcomers arrived from across the sea and encountered this bird they had never known, they also apprehended its strength. Perhaps it was just anthropomorphic projection. Bald eagles mate for life and carefully tend their young, and maybe the settlers imagined some kind of puritanical family values. Bald eagles also have a passion for building—their nests can weigh up to a ton and last for decades—and maybe the restless settlers appreciated the bird's industriousness. At the very least, the new Americans liked the bird's military bearing, with its steely gaze, broad chest, and bone-white crown. In 1782, the bald eagle became the central symbol of the Great Seal of the United States, the unofficial national bird—George Washington in feathers.

But it was a conflicted relationship. The eagle had become a symbol of America, yet many Americans didn't like the actual animal. "He is a Bird of bad moral Character," Benjamin Franklin famously wrote. "He does not get his Living honestly."

Franklin was referring to the bald eagle's habit of stealing fish from the smaller osprey, its penchant for poaching, and its taste for carrion. Settlers complained that eagles often snatched their lambs and chicks. Against all common sense, the myth spread that bald eagles could carry off human babies from the crib.

A campaign of ecocide began. In the nineteenth and early twentieth centuries, Americans murdered eagles mercilessly. Farmers and ranchers shot them on sight—"We kill all the eagles we can," one sheep farmer said—while some state and local governments paid bounties for dead eagles. In Alaska alone, hunters killed some eighty thousand eagles to cash in on the territory's bounty. By the turn of the last century, the birds were vanishingly rare in most of the country. "Twenty years ago, I used to see a dozen or more along the river in the spring" the nature writer John Burroughs noticed in the early 1900s, "where I now see only one or two, or none at all."

Eventually, the American public realized the country's mistake. In 1940 the Bald and Golden Eagle Protection Act outlawed killing the bird. But the protection came almost too late. A new danger worse than bullets and buckshot threatened the species: the powerful insecticide DDT, which government agencies and farmers were spraying all over the place. The bald eagle—along with osprey, peregrine falcons, and pelicans—became collateral damage in the war against insects. DDT weakened the eagles' eggshells and lowered females' hormone levels, impacting the species' ability to procreate. In *Silent Spring*, Rachel Carson wrote, "Something is at work in the eagle's environment which has virtually destroyed its ability to reproduce. . . . The reports differ in detail, but always repeat the theme of death." By 1963, there were just 417 nesting pairs outside of Alaska.

In a spate of legislative urgency that these days feels like something out of legend, lawmakers rushed to address the crisis. Congress banned DDT, which was pulled off the market in 1972. When the Endangered Species Act became law in 1973, the bald eagle was among the first species to come under its protective umbrella. The Clean Water Act helped too. An eagle can't survive if it has nothing to eat, and by the 1960s many of the United States' rivers served "as little more than sewers to the sea," in the words of one senator. The Clean Water Act prohibited industries from using streams, rivers, and lakes as dumps for their chemical crap. Freed from pollution, the waters slowly came back to life. Fresh waters again furnished the food web.

The descendants of the destroyers became the restorers.

State fish and wildlife agencies launched captive breeding programs in which biologists, often assisted by volunteers, pantomimed the behavior of eagle parents to raise chicks. Using a technique called "hacking" that was already common among keepers of trained falcons, biologists reintroduced young eagles to lakes and rivers. Citizen scientists kept vigil over nests and helped count bald eagle numbers. Popular attitudes changed. For many Americans, to kill a bald eagle came to be something akin to burning the flag.

The heroic labor paid off. In 2007, the bald eagle was removed from the endangered species list. Between 2009 and 2021, the number of bald eagles in the Lower 48 quadrupled, from about 72,000 to more than 300,000—a figure not all that far from the estimated baseline before European arrival in the Americas. In the space of a couple human generations, the nearly extinct became common again.

It was as if the great hoop of time had spun all the way around. *Once*—that word that weighs the scales of every shifting baseline—could be used in celebration instead of lamentation.

After an absence that stretched beyond memory, the bald eagles arrived at Madison's Warner Park in the summer of 2020. Warner Park is one of the largest parks in the Wisconsin state capital, one of those jewel boxes of city nature that—like New York's Central Park or Audubon Park in New Orleans—forms an ecotone where the urban, the cultivated, and the undomesticated meet. Braceleted by the tracks of the Wisconsin & Southern Railroad, encircled by modest middle-class neighborhoods, and just a five-minute drive from the local airport, Warner Park is no wilderness. But it's still a wild place, home to beaver and muskrat, deer and fox. An unruly copse—locals call it "the Big Thicket"—is packed with red maple, ash, and walnut. A massive bur oak, some 250 years old, has lived there since well before the first settlers appeared. Warner Park is one of those spots where people can watch wildlife and the wildlife are just comfortable enough with humans to watch back.

For the eagles, Warner Park had what they needed to make a

home. When a mated pair began building a nest on an island in the park's cattail-ringed lagoon, neighbors quickly spotted them. The birds became minor celebrities, symbols of hope in that eerie season of pandemic lockdown. "It was this brilliant beacon of light during a very dark time," Beth Sluys, a Warner Park neighbor and amateur naturalist who had long been involved in local conservation issues, told me. "Observing the birds living their lives, there's something magical about that."

A neighborhood teenager named Elsie Olmanson who had started watching the eagles with her father after hearing about them from Beth also found joy in their arrival. "You're happy that they are able to have a recovery, and that they're coming more and more back into the world," she told me. The eagles, she said, were "adorable creatures." The nesting eagles seemed to her "like a married couple—they will bicker with each other."

By the time of autumn leaf fall, the nest was impossible to miss: a lump of sticks and branches some thirty feet up, the whole mess wedged in the crook of two limbs, like a leather sofa had been dropped from the sky. The calendar flipped to 2021, and that March the bald eagle couple hatched a single eaglet. At which point a minor controversy broke out. Just a few hundred yards from the eagles' home lagoon sits "the Duck Pond"—a small stadium that is home to the Madison Mallards, a collegiate summer baseball team. Every Fourth of July the Mallards stage a big fireworks display, and some of the eagle fans feared the fireworks would scare away the birds. A few folks petitioned the baseball team to halt the pyrotechnics or, at the very least, postpone them until after fledging time. "We were concerned that this might potentially impact the fledglings at a critical time," Elsie's father, Eric, said.

Other eagle enthusiasts were nonplussed. Patrick Hasburgh, a passionate outdoorsman who owns a fishing bait and tackle shop on the edge of the park, was among the first people to spot the nesting eagles, and he felt like the fireworks controversy was overblown. "I've been watching the nest since it was built," he told me, "and the eagles don't seem to mind [human activity] at all. They are used to people. It's an urban environment."

In an age of mass extinction, the dustup over whether a fireworks show would disturb an eagle family seemed to me like something of a First World problem. Fifty years ago, there were just 108 eagle nests in all of Wisconsin; by the time an eagle pair started stacking sticks in Warner Park, that number had climbed to nearly 1,700. With eagle survival no longer in doubt, the question had become how best humans might help eagles to thrive. Here was another form of shifting baseline syndrome—an example of how progress, every bit as much destruction and diminishment, can be difficult to remember.

Too much of environmental history is a litany of losses. Ghost stories haunt the landscape: the passenger pigeon, the Tasmanian tiger, the ivory-billed woodpecker. But running parallel to the string of disappearances are stories of recovery and resilience. The bald eagle is not alone.

Think about the forests of North America: there's more forest in New England today than there was in 1850. Consider the air and the water: both are clearer now than before the Clean Air Act and Clean Water Act were passed. Recall the beaver—now numbering in the millions across North America—or loggerhead turtles, whose population increased by nearly two orders of magnitude in the last twenty-some years. The California condor, at one point on the doorstep of extinction, is on the wing again. Take a breath and remember the whales, saved from annihilation.

Seen from one vantage, many of those success stories could also be evidence of shifting baselines, of how our ideas of abundance contract generation by generation. Sure, there are more wolves on the American landscape now than there were fifty years ago—and there are also far fewer wolves than there were 150 years ago. But Mia Monroe, the co-founder of the Western Monarch Count, often brings up such wildlife recoveries as examples of the possibilities that also exist for her beloved butterflies. If we could save the bald eagle, what else can we save? "In my life I've watched the gray whale return and the peregrine falcon and the otter," Mia told me once. "I don't know what's ahead for monarchs. But I do know that when people realized there was a problem, and we did something different, then nature responded. I always remember those examples."

I remember them, too. They remind me that baselines sometimes shift in virtuous directions. The British writer Robert Macfarlane suggests a name for this: *lifting baselines*. "We can normalize betterment as well as decline and then wish for betterment's increase," he has written. "We can run the reel backward." The bald eagle's comeback is proof that the passage of time doesn't always mean loss and decay. Growth and increase are also expressions of time. So is healing.

There is, however, a danger that can come with lifting baselines and the normalization of betterment. To grow accustomed to abundance risks obscuring all the hard work and long years it took to achieve that abundance. Without collective memory, improvements can also be difficult to see. Rebecca Solnit made a similar point in an essay a few years ago when she warned that "forgetting is everywhere." Solnit wrote that she feels like "a tortoise at the mayfly party," and cautioned that "change itself becomes invisible when your timeframe is shorter than that change." The indivisible present can cloak the march of progress just like it provides cover for the grind of destruction.

Environmental progress is especially vulnerable to becoming invisible since, more often than not, it leaves no trace. Many environmental victories are triumphs of harm avoided: the proposed oil pipeline denied, the forest kept free from the loggers, the fossil fuels unburned. Pollution, once cleaned from the air and the water, is another one of those absences that the mind struggles to recognize and remember. Without memory, environmental well-being is always at risk of being taken for granted. To endure, progress requires vigilance—the long view of history.

When I was growing up, the bald eagle was a rarity; the first time I saw one, I was in my mid-twenties. Now, I live in a place where bald eagles are common to the point of mundane. I make an annual winter-solstice pilgrimage to the North Fork of the

Nooksack River, where bald eagles stage big convocations. From the small bridge there I can spot scores of eagles perched in the December-bare alder and cottonwood, their calls—*skreee*, like a door on rusty hinges—echoing off snow and ice. It's a thrill to watch them scrimmage over salmon carcasses. For me, an eagle sighting never gets old.

My daughter has a different relationship to the bald eagle, especially having grown up in a place where you can spot them at the county dump (*He does not get his Living honestly*). She barely registers them as special, and she all too often greets her dad's glee with an eye roll. What is a bald eagle to a child who has never known them as endangered? About as interesting as a wren.

I got a welcome corrective to these worries after spending a morning with the family of Patrick Hasburgh, the bait store owner who has been involved with the Southern Wisconsin Bird Alliance's formal Bald Eagle Nest Watch since the mated pair showed up Warner Park. The Hasburgh boys—Fletcher, then sixteen, and Ranger, eleven—had spent countless hours with their parents monitoring the eagles. But the commonplace had not dulled excitement. If anything, their regular eagle-watching had made them fascinated with the birds—appreciation earned through intimacy. Fletcher told me he "absolutely" knew the story of the fall and rise of the bald eagle, in part "because I have, you know, a kind of field knowledge." Ranger, who was sporting a skater's cut and an Off-the-Wall black hoodie, said he liked to "see the birds up close and see their whole life cycle." He liked how, when he uses his dad's spotting scope to check out the nest, he can "pretty much see every individual feather."

We strolled the park, and the family shared with me some of its natural wonders: a killdeer nest on the railroad tracks, a pair of sandhill cranes coming in for a landing. The birds were like giraffes on the wing, tumbling earthward and honking madly. It quickly became apparent that the Hasburghs were involved in all sorts of neighborhood-scale environmental stewardship projects. Here was an ordinary American family who intuitively, organically, was embodying the antidotes to environmental amnesia—going outside in all weather and all seasons, emotionally encoding their relationships with the living world, writing them down, and transmitting their experiences back and forth between generations. This focus group of four seemed the very model of Emerson's "practical naturalists."

As a family, they go outside all the time. When Patrick and his wife, a nurse practitioner named Ashley, started a family, they committed to get their kids outdoors as much as possible because, as Patrick said, "they have more than enough reasons to stay inside." The family made a point to visit every state park in Wisconsin. They hiked a lot, fished a lot. And they adopted Warner Park as their backyard wilderness. "This park has been a constant through my life," mop-headed Fletcher said as we walked the paths not far from the eagle nest. "Now, when I'm going through this park, there's a million stories."

Over time, the family attuned themselves to the urban wildlife, began to notice things that other people seemed to miss:

the cranes and the eagles, tiny frogs in the lagoon, a big snapping turtle laying its eggs on the railroad tracks. "For me, it's one of those things where it's like: Is it that I'm more aware, or are things changing?" Ashley said. "I see more nature than I remember seeing when we first started walking. . . . Maybe I'm more in tune, or maybe it's that it's actually just kind of started becoming more wild. Or both. Probably both."

The Hasburghs had made a record of their experiences in Warner Park. In addition to the eagle nest monitoring, Patrick enlisted the family in an initiative to protect the area's bluebirds, believing that a formal citizen-science project would be a way of "sneaking some education in . . . without it being their lame dad preaching to them." On a weekly basis the family checked on eleven bluebird nesting boxes, evicted any sparrow squatters if necessary, and then sent to the Bluebird Restoration Association of Wisconsin information about the number of eggs and chicks.

When he was a little kid, Fletcher had taken the work very seriously. Equipped with a clipboard, he studiously went from box to box making observations. "Pretty much half of my entire summer was spent just checking bluebird boxes," he told me as we inspected one nest box that, it turned out, was empty. By now, Ashley said, the bluebird stewardship was "built into our lives."

Both boys had become crackerjack birders who could tell a sparrow nest from a chickadee nest from a bluebird nest at a distance. They could see a bluebird chick and tell how many days old it was. With familiarity came emotional attachment: the family imprinted on the birds. "Maybe it's just in my head, but you build sort of a relationship with them," Ashley said. "I love it, and it's fun to see, like, the evolution. And you kind of feel like, you know, a little mama bird."

Passing on their naturalist knowledge to other kids was a bigger lift. Fletcher—whose great passion is video games, especially *Call of Duty*—said he didn't exactly advertise his conservation endeavors. "I don't really want to have a big prom sign that says, *Come to the prom with me—I look at birds.*" And as much as Ranger was psyched about outdoor activities like fishing—the previous winter, he had caught a "monster" thirty-eight-inch pike and a twenty-nine-inch walleye—most of his free time was spent skateboarding.

If anything, the boys had passed their environmental passion not outward to peers but upward to their parents. Patrick said one of the reasons he had gotten into birds is because Fletcher, when he was very young, kept asking for their names. "And I wouldn't have an answer," Patrick said. "So then I got curious and then we started looking at other birds and we started a life list." A child's inquisitiveness, it turned out, could be a parent's spark bird.

When we first spoke, Patrick told me that one his great inspirations was Aldo Leopold, whose writings he had been exposed to during a brief stint in college and who moved him to seek out "an engagement with the landscape." It was *A Sand County Almanac* that spurred Patrick to start keeping a fishing journal, which turned into a birding journal, and then into an all-around phenology journal that is now entering its twenty-fifth year. "It's mostly birds and animals and stuff like that," he told me. "And at the end of the year I go through my journals and notes and send out a little report to friends."

A famous line in Leopold's canonical book comes as the author declares, "One of the penalties of an ecological education is that one lives alone in a world of wounds. Much of the damage inflicted on the land is quite invisible to laymen." The Hasburghs' tour of Warner Park made me think that maybe it was time to turn Leopold's insight onto its head. Perhaps it's equally true that one of the *benefits* of an ecological education is that one might, together with friends and family, inhabit a world of wonders. Much of the abundance that remains on the land is made visible by sharing it.

Isn't this how the world might turn toward environmental progress, one workaday relationship with nature after another? The Hasburgh family was living proof that the task of ecological remembering isn't necessarily difficult or esoteric; it can be as simple and as easy as a walk in the park. And as much as the Hasburghs demonstrated *how* to make memories of the living world, they also illustrated *why* it's worthwhile to do all this physical, mental, and emotional work. Remembering isn't about dwelling in the past, but a way of casting oneself forward in time, into the future tense. It was Ranger who helped make this clear. "I like just seeing the eagles and just watching the nest," he said as we wrapped up our loop around the park and stood, again, not far from the eagle eyrie "That's, like, a memory that will stay with me for a while."

So, then: Can we prove ourselves smarter than a frog in hot water?

After all the explorations, it seems to me that the best chance of passing the existential IQ test hinges on one of our species' most uncanny skills: the ability to time travel, to mentally shuttle between past, present, and future. Cynics willfully misunderstand the endeavors of environmental conservation as merely backward-looking, some kind of fetish to preserve a forever-lost golden age. But even this fallen Eden deserves to be protected. Today, after all, is the starting point of someone else's ecological baseline; we are descendants and ancestors all at once. Making that empathic jump relies on the human capacity for imagination—a power that itself rests on the ability to remember. Memory is the axis that connects yesterday to tomorrow.

"It's a poor sort of memory that works only backwards," the White Queen says to Alice in Lewis Carroll's *Through the Looking Glass*. Thankfully, memory doesn't work that way. As often as not, memory is a way of catapulting into the realm of imagination. Just as memory is a refashioning more than an exact retrieval, imagination is a recombination of things already known and experienced. Our imaginings rest on remembering. And there, at the junction of memory and imagination, resides our best chance of being good future ancestors, of bequeathing a world more vibrant than the one we've inherited.

Scientists have long guessed at a connection between remembering and imagining. Nearly a hundred years ago, a pioneering English psychologist, Frederic Bartlett, wrote, "Remembering is not the re-excitation of innumerable fixed, lifeless and fragmentary traces. It is an imaginative reconstruction." We don't directly recall the past—rather, we reimagine it. Every episodic memory is an act of conjuring.

But it wasn't until this century that neuroscientists confirmed a biological link between imagination and memory—a discovery so profound that *Science* magazine called it one of the top ten scientific breakthroughs of the year. By studying people with a damaged hippocampus, researchers confirmed that people with severe amnesia find it close to impossible to envision future scenarios. Then neuroscientists using brain scans established that a common neural network underlies memory and imagination. The scans of healthy brains revealed that the hippocampus is even more engaged when subjects visualize the future than when they recall the past.

From an evolutionary standpoint, one of the main reasons we remember things is to allow us to navigate life's choice points. "A key function of memory is to provide a basis for predicting the future," one group of researchers has written. Past experiences—mapping where the game herds migrate, where the tastiest fruit is found, where the dangerous animals lurk—inform crucial decisions about the future. Memories of something that actually happened provide the foundation for envisioning what might or could happen. The purpose of memory is imagination.

Even fantasy is made from the stuff of real life, recast and then tuned up. To picture a Pegasus, one needn't have seen an actual flying horse. It's enough to have seen a real horse and a real bird and then combine the two. Memory provides the yeast for imagination's fermentation.

Poets long intuited this. In his poem *The Prelude*, the English Romantic William Wordsworth identified "spots of time"—evocative flashbulb memories—as a wellspring of creativity. Through the "renovating virtue" of memories, "our minds are nourished and invisibly repaired," enhancing one's imaginative powers. "Such moments are scattered everywhere, taking their date from our first childhood."

Historians, naturally, have also made the connection between memory and imagination. The collective inter-generational memories that make up the corpus of history do more than memorialize the past. They also expand the prospects of the future. "We know these things about the future only from having learned about the past," historian John Lewis Gaddis writes in *The Landscape of History*. "History is arguably the best method for enlarging experience." As the time machine of memory enlarges experience, it provides new perspectives, fresh ideas, radical notions even. History unbinds the imagination.

Social-change activists know better than anyone the ways in which memory preserves possibilities, how forgetting constricts the terrain of the future. "The short-term view breeds defeatism and despair," Solnit, a people's historian if there ever was one, writes. "If people are shortsighted about the past, so they are about the future." What is to be done? "The most important territory to take is in the imagination," she declares. "Once you create a new idea of what is possible and acceptable, the seeds are planted."

Imagination is the alpha and the omega of political action. But without memory imagination is handicapped, leaving us stranded in today's status quo. The realm of the possible is demarcated by the boundaries of expectation—which are themselves delineated by how little we remember and how much we forget. So: the better we remember, the greater the borders of possibility. From memory is the cartography of imagination made.

For the dissident—and, let's face it, anyone who works to protect the planet in the era of late capitalism is a dissident to some degree—a long memory shows that progress rarely moves in a straight line. Progress zigs and zags, serpentines like a wild and free river. All sorts of routes ahead are possible. And even as memory safeguards potential and possibility, it unlocks the impossible by illustrating how many impossibilities have already come to pass: the whales saved from utter destruction, the bison rescued just in the nick of time.

These ideas about the relationship between past and future—about the intersection of memory and imagination—are as important today as they have ever been. Because shifting baseline syndrome isn't merely a cognitive maladaptation or a psychological coping mechanism. Amnesia is also a political agenda. Power, after all, always seeks to shape what seems normal. While some may wear the blinders of forgetfulness to try and get by, many others have been deliberately blinded.

There are people working right now to foment forgetting. Here in the United States, wannabe authoritarians are waging a war on history and memory. Historical plaques at national parks, national monuments, and battlefields are being rewritten to change any narratives that might reflect negatively on the nation's past. The Smithsonian Institution—that great repository of American history—is facing government officials' threats to revise the stories it tells. The federal government is removing from official websites certain words that are now considered verboten. Forbidden words include "advocacy," "oppression," "Native American," "climate change," and even "female." Also on the list: "historically."

This weaponization of natural human forgetfulness—organized forgetting, to return to Milan Kundera's piquant phrase—has long been encouraged by the forces of environmental exploitation. At some level, the corporate functionaries of the extractive industries understand that environmental amnesia enables their agenda as much as it obscures it. See, for example, the web of disinformation woven by the fossil fuel giants.

After spending decades denying the science of global warming, companies like ExxonMobil now deny that such denial ever took place—an attempt to memory-hole their well-documented campaigns of misinformation. Each year Big Oil spends hundreds of millions of dollars, if not billions, on advertising and marketing, including ad buys that seek to craft narratives about the supposed necessity of society's dependence on oil and gas. "Oil pumps life," we've been told. "Energy in progress," Chevron trumpets. According to academics who have decoded fossil fuel messaging, there are scores of examples of oil-company advertising that insist on the inevitability of climate change and/or an indefinite reliance on fossil fuels. Such narratives, they report, "normalize fossil fuel lock-in." It's a storytelling sleight of hand—corporate gaslighting in the form of slick PR—that makes the status quo appear inalterable. What is "fossil fuel lock-in" if not an attempt to imprison society in an invisible present, ever-dependent on the energy sources that are burning up the planet?

The prescriptions for environmental memory that I've sketched—the creation of personal ecologies, the cultivation of emotions, the crafting of records, the transmission of ecological histories—are antibodies against the virus of organized forgetting. Such actions help open the door to alternative futures beyond the profit calculations of corporations or the fever dreams of fascists. All of which makes the seemingly simple act of memory nothing less than a subversion of business as usual. Remembering is a form of resistance.

The great danger of shifting baseline syndrome is the way in which it narrows our expectations and imaginings, does so by compressing the definition of normal. A child who has only known the pollution of an oil refinery neighborhood empty even of birdlife will understandably expect that's the norm. A kid who has only experienced summers of lethal heat might reasonably assume the climate has always been that way. If she's never heard family members' personal stories about the great forests or the life-soaked seas, the old-growth groves and an ocean full of fish might as well be fantasia.

On the flip side, to remember that *today* is not synonymous with *forever* widens the horizon of the possible. It's easier to picture the coastal California pines again cloaked in orange and black if you've heard the old-timers' tales about how they used to be flocked by millions of monarchs. Black-and-white photos of the salmon in massive runs enable one to visualize the rivers running full once more. Knowing the legend of the bald eagle's revival is an inspiration that such a thing could be repeated. Our ecological imagination will only be as fertile and as wild as our remembrances of Earths past.

How many of our ecological aspirations rest on the cornerstone of memory! We hope to restore ecosystems, rehabilitate waterways, revive scarred landscapes, rewild the world however best we can. Renewal, rejuvenation, reassemble, reclaim, recreate, reunion. Consider the etymology: *Re* means "back from" or "back toward the starting point." The most resilient environmental memories, as I insisted earlier, will be found in the rediscovery that ecological time works more like a circle than a line. In a spiral, every return is also forward motion.

What's at stake in this work of remembering and imagining is nothing less than the chance to flourish. No matter how awful things might get, *Homo sapiens* will almost certainly habituate and accommodate. As Dostoyevsky wrote, we are creatures that can get used to anything. All those neighborhoods crowded against the oil refinery fence lines—all those homes incinerated in climate change–fueled fires—prove as much. But we can do better than merely adapt and muddle through. By forging memories into imaginings, we give ourselves a better chance of not merely surviving but truly thriving.

Envision it: true abundance for all, for both humans and nonhumans, all those other beings with whom we've evolved to arrive at this place in time. A civilization whose affluence is not mortgaged on poisonous ancient sunlight that must be torn from the earth, but rather one whose well-being comes from the energy of today's sun. Human society cohabitating in rough equilibrium with the landscape. The skies jammed with birds again.

To have a long memory is to know that such dreams are possible. Anyone who has heard some Native lore can picture what it would be like to live in neighborly grace with the more-than-human. Anyone who has read the records of natural history knows about the natural wealth that was once considered normal. When we pierce the veil of the invisible present, we give ourselves—and all the other life forms with whom we share the planet—a better chance of thriving. Remembering is part of the work of flourishing.

Even in this age of discontinuity, when the past seems a poor instructor for an unprecedented epoch, memory can serve as a guide. It can be a compass pointing the way forward. By doing the work of going outside, bearing witness, making a record, and then passing it on, we will discover whole new geographies on this singular planet. At the crossroads of memory and imagination we will find ourselves in a landscape of wide-open possibilities.

NOTES

"In order to write so much as a sonnet, you need a warehouse," Annie Dillard once wrote. To write an entire book, then, might require something approaching a city block of mega warehouses.

The full citations for this book stretch to twenty thousand words, a length that, were it to be printed in full, would threaten to eclipse the main text itself. So I have placed most of my supporting evidence online. Scholars and the insatiably curious can find the complete citations at jasondovemark.com/the-earth-said-remember-me-notes.

The notes here are amplifications, detours, and asides for extra edification and entertainment.

INTRODUCTION: FROG IN HOT WATER

5 **the public response to the tentative rebound:** When I started tagging along with some of the monarch counters during the tail end of the 2022 Thanksgiving Count, I was struck by how the idea of shifting baseline syndrome was already well known within that community. Emma Pelton at the Xerces Society had brought up shifting baseline syndrome in an interview with *Time* (see: Oliver Millman, "The Fight to Save the Monarch Butterfly," March 21, 2022). In a post on their LinkedIn page, the staff at the Pacific Grove Museum

of Natural History had introduced the concept to their docents and supporters: "People think of how things used to be in their childhood as the golden period. As things progressively get worse, new generations can't imagine a better world, thus making it harder for conservation groups to work towards a healthy ecosystem. . . . The goal is not to equal the number of butterflies we saw in 2021, to be more than 2020, or even to reach 1997. The true goal is the historic tens of millions that used to exist across all of California."

7 **the steady coarsening of American civic and political culture:** In a *New Yorker* article exploring whether democracy in the United States is sliding toward demise, Andrew Marantz makes a comparison between how the public has slowly accommodated to global climate change and the American political system's slide toward authoritarianism. "In the real world, though, the cataclysm can come in on little cat feet. The tremors can be so muffled and distant that people continually adapt, explaining away the anomalies. You can live through the big one, it turns out, and still go on acting as if—still go on *feeling* as if—the big one is not yet here." See, "Is It Happening Here?" *The New Yorker*, April 28, 2025.

8 **The ancient wisdom is right:** The insight about how you can never step in the same river twice comes from the ancient Greek philosopher Heraclitus. Brooke Jarvis mentions Heraclitus in her groundbreaking article about the so-called insect apocalypse. "It is impossible to maintain a fixed perspective," Jarvis wrote. See "The Insect Apocalypse Is Here," *New York Times Magazine*, November 27, 2018.

9 **"Man is a creature that":** The theme of accommodation as a survival technique runs through Dostoyevsky's novel. At another point in the book, he writes: "Working under escort in the barracks together with two hundred 'companions': never alone, never! However, I was obliged to get accustomed to it."

10 **The concept of shifting baseline syndrome was coined:** Pauly's article, "Anecdotes and the Shifting Baseline Syndrome of Fisheries," has been so influential because of its intrinsic insightfulness and prescience. But what I find most interesting is how the word "anecdote" slipped into the title. As I'll explore, anecdote-driven community science and grassroots oral history and natural history are some of the

most important antidotes to environmental amnesia. Speaking of amnesia, Pauly never uses that term in his original paper, but the focus on memory does appear in his subsequent writings as he dissects the idea of "collective amnesia." See Pauly's 2019 book, *Vanishing Fish: Shifting Baselines and the Future of Global Fisheries*.

10 **shifting baseline syndrome has been referenced:** According to the National Institutes of Health research database, the paper has been cited some 3,400 times. In their paper, "The Shifting Baseline Syndrome as a Connective Concept for More Informed and Just Responses to Global Environmental Change" (*People and Nature*, May 2023), Heidi K. Alleway et al. say the paper has been cited 1,560 times. I'm splitting the difference. In any case, as Alleway and collaborators report, "This is an impressive reach for a one-page article." The authors uncover the historical nugget that an Australian fisheries expert, H. C. Dannevig, coined the term "drifting baselines" way back in 1903.

10 **Should we be trying to restore:** In recent decades, shifting baseline syndrome has become one of the knottier questions in conservation biology. To what baseline are we hoping to restore Wetland X or Grassland Y? It's an epistemological problem: biologists may not know exactly how things were in the past, especially the distant past. A perfect distillation of the dilemma comes from Scott D. Sampson and Peter D. Roopmarine, "We Need to Think About Conservation on Different Timescales," *Scientific American*, October 15, 2023, in which they write: "Time is one of humanity's greatest blind spots. . . . Humanity's shortsightedness around time creates major constraints on modern conservation. As the climate and biodiversity crises accelerate, we are urgently working to protect and regenerate ecosystems without understanding how they functioned when they were truly thriving."

13 **the concept has trickled into the public dialogue:** For several years I've been receiving Google Alerts for "shifting baseline," and I receive, on average, about two alerts a week. Shifting baseline syndrome has become sort of like the Forrest Gump of environmental discourse, constantly popping up here and there, sometimes in the foreground, sometimes in the background. In the preface to the latest edition of

her powerful book, *On Extinction*, Melanie Challenger writes, "Little wonder, then, that the idea of shifting baselines has become such a hot topic in research. . . . This generation is then less able to judge how destructively altered are the places they inhabit, and so it goes on."

14 **The world won't end with the bang of calamity:** The critic Amanda Hess investigated this phenomenon some years ago when she wrote, "I can't say precisely when the end began, just that in the past several years, 'the end of the world' stopped referring to a future cataclysmic event and started to describe our present situation. . . . Ours is a banal sort of apocalypse. Even as it is described as frightfully close, it is held at a cynical distance." See "Apocalypse When? Global Warming's Endless Scroll," *New York Times*, February 3, 2022.

15 **This rejiggering of weather recordkeeping:** NOAA's 2021 release of the updated thirty-year averages sparked a burst of news articles about "the new normal" and shifting baseline syndrome. Veteran environmental journalist Bryan Walsh anticipated the handwringing in his dispatch, "Shifting Baselines Are Changing What Normal Means," *Axios*, January 6, 2021, when he wrote, "Some climate scientists have urged weather agencies to fight shifting baselines by sticking with a static 30-year time period when calculating weather averages, rather than updating the reference period every decade."

16 **Michael Mann, an eminent climatologist:** Long after the new figures had been released, Mann's righteous annoyance was undimmed. I emailed him in spring 2022 and asked for his comment, and he replied: "When we redefine 'normal' as 'today,' it literally 'normalizes' global warming. It's a continually shifting baseline toward more extreme and damaging impacts. I know how dramatically things have changed from my own lived experience. But the students I teach at Penn State don't have that experience. A warmer world with more damaging weather extremes, coastal inundation, etc. is what they've grown up with. So they don't have the same profound, intuitive sense of how dramatically things have changed."

17 **unusual temperature variations cease to spark interest:** See: Frances C. Moore et al., "Rapidly Declining Remarkability of Temperature Anomalies May Obscure Public Perception of Climate Change," *Proceedings of the National Academy of Sciences*, February 2019. The authors

write, "Human evaluation of weather as normal or abnormal will also be influenced by a range of factors including expectations, memory limitations, and cognitive biases." This is probably one of the most frightening shifting baseline syndrome studies out there.

17 **I put my body against the proverbial gears:** I spent much of my twenties working for a scrappy human rights group called Global Exchange, where, among other endeavors, I helped launch a national campaign targeting the American automakers and urging them to break the US addiction to oil. This was at the height of public opposition to the US invasion of Iraq, and our theory of change (as advocates say) was to take the "No Blood for Oil" slogan of peace activists and move it upstream to the auto companies. I was twice arrested during that campaign: once for hanging a banner off a Ford dealership in San Francisco, and then for being involved in another banner hang during the LA Auto Show.

18 **many people believed that global civilization's laggard response:** If you read any environmental journalism in the 2000s and twenty-teens, you might recall frequently hearing this argument, which is still coasting on vapors today. For example, see a 2018 video, "Why Humans Are So Bad at Thinking about Climate Change," produced by *Vox* and the University of California, which includes environmental thought leaders like M. Sanjayan of Conservation International and Anthony Leiserowitz of the Yale Program on Climate Change Communication.

20 **Even if you're just hearing:** In a TEDx talk about shifting baseline syndrome, Pauly quips, "The idea can be explained in one minute."

20 **if you put a frog into a pot of hot water:** The metaphor was made famous, in part, by Al Gore's Academy Award–winning documentary, *An Inconvenient Truth*. A good indicator of the persistence of this trope is the fact that when two of the authors of the *PNAS* study about social media and extreme weather wrote up their findings for a popular audience, they felt compelled to mention it. See: Nick Obradovich and Frances C. Moore, "The Data Is In. Frogs Don't Boil. But We Might," *Washington Post*, February 25, 2019.

20 **urban myths that just won't quit:** Famously, a lie speeds around the world before the truth has its pants on. The trope has been repeatedly

debunked, yet it persists. Victor H. Hutchison, a herpetologist and professor emeritus of zoology at the University of Oklahoma, appears to have actually run the experiment back in 2002. The "bullshit" line comes via a delightful takedown by James Fallows, "The Boiled-Frog Myth. Stop the Lying Now!" *The Atlantic*, September 16, 2006.

21 **focus on the antidotes to landscape amnesia:** The term "landscape amnesia" appears to have been first used by geographer Jared Diamond in his book, *Collapse* (pages 425–26). Diamond coins the term in the context of the apocalyptic ecological history of Easter Island—a forested paradise that was clear-cut and denuded so severely that the Polynesians there eventually resorted to cannibalism. We were warned.

CHAPTER 1: GO OUTSIDE

25 ***remember*, you might say:** In *Searching for Memory: The Brain, the Mind, the Past*, psychologist Daniel L. Schacter makes an important distinction between knowing and remembering (see pages 22–26). You may know you were at the dinner with Joe and Jane, but do you really *remember* being there? This distinction is key when it comes to embodied memories. It's not enough to know that, say, white trillium blooms in the early spring. Actual visceral memories are required to overcome landscape amnesia.

27 **Some place cool, green, and watered:** I wrote more fully about the experience of fleeing California in part over climate concerns in the essay "I'm an American Climate Emigrant," *Sierra*, October 12, 2020. "It wasn't really the fires, though, that drove me from California, I wrote. "No, it was the specter of drought—the fires' flinty handmaiden."

27 **moss swaddling limbs:** There are some 22,000 species of moss worldwide, but as Robin Wall Kimmerer writes in *Gathering Moss*, they "don't usually have common names, for no one has bothered with them."

27 **I bought a pile of field guides:** My favorite for this bioregion is *Cascadia Revealed* by Daniel Mathews, who is authoritative and also not above the occasional dad joke. The ethnobotany is first-rate.

27 **made birdsong recordings with Merlin:** Some environmentalists worry that the proliferation of nature apps is just further distancing people from nature, another form of technology mediation. "Nature apps might undo . . . ways of knowing nature that are based on observation, experience, and skill. In fact, that is the whole point of most nature apps: The user can get answers without having to do much," writes Paul Keeling in "Time to Rethink Nature Apps," *Sierra*, November 29, 2022. Maybe. But these technologies have helped make naturalist knowledge more democratic. I, for one, would rather err on the side of accessibility versus purity.

28 **citizens' efforts to preserve the living world:** You'll note the phrase "living world" a lot in these pages. It is, in my opinion, one of the best synonyms for the overused and often misunderstood "nature."

30 **His memoir, *The Thunder Tree*:** Taken as a whole, Pyle's book is a celebration of the work of developing a sense of place. In the prologue he writes, "When people connect with nature, it happens *somewhere.*"

32 **All generational amnesia begins with individual forgetting:** One of the scariest examples is the now-infamous episode in which the *Oxford Junior Dictionary* in 2007 and then again in 2012 nixed dozens of nature-related words to make room for other words connected to modern technology and communication. The removal of words like *almond*, *crocus*, *catkin*, and *chestnut* prompted protest from writers including Margaret Atwood and Robert Macfarlane, who worried that a "place literacy is leaving us." Language deficit has compounded nature deficit. As neuroscientist Lisa Genova notes, "Without language, there is no memory."

34 **Do you know the names of the trees:** This provocative question comes from no less an adventurer than Steve Casimiro, the editor of *Adventure Journal*, who posed similar questions to his ultra-outdoorsy readers in his Spring 2025 editorial.

37 **highly recommended for forestalling dementia:** My interest in neuroscience isn't just an academic curiosity. About ten years ago, my father was diagnosed with cognitive impairment disorder, which then bloomed into dementia, and finally into full-blown Alzheimer's disease. It's been an awful experience—and, at the same time, a revelatory one. To witness my old man disappear before my eyes has pro-

vided a window into how essential memories are to our being. We are made of our memories, every bit as much as a house is made from concrete and wood. Memory is the cornerstone and keystone of individuality. Eventually, someone suffering from Alzheimer's becomes stranded in a purgatory of the permanent present. What is true for individuals is also true for cultures: an inability to remember leads to a loss of self.

37 **an area of the brain called the hippocampus:** It's complicated; memory functions are distributed throughout the brain, and it's partially this multi-nodal processing that makes memory an associative exercise. And still, as Charan Ranganath writes in *Why We Remember* (page 36), "To many neuroscientists, the hippocampus is synonymous with memory."

37 **This overlap between memory and spatial-temporal orienteering:** Most of this book was written while walking. I always carry with me a hip-pocket notebook that I use to jot down ideas whenever they may come to me. Most often they arrive when I'm out hiking in the woods with my dog. The notes eventually make their way to the keyboard and onto the page when I sit down to put words and thoughts together—which is to say, to do the work of writing. But, truly, most of what I consider the better lines and most original ideas of this book came to mind while I was ambulatory.

38 **an "archive of the feet":** Schama is a cultural historian, and so it's no surprise that he views nature through the prism of culture. "Landscapes are culture before they are nature," he writes on page 61 of *Landscape and Memory*, "constructs of the imagination projected onto wood and water and rock." Well, no—the wood is wood and the water is water and the rock is rock. Even if it's true that we humans can only experience nature through the lens of our own perceptions, the raw material of Earth exists separate from the human mind. A tree crashing in the woods makes a boom even if no one hears it.

38 **multisensory memories that combine a visual cue and an auditory cue:** It's significant that many people with exceptional memories, like the Russian mnemonist Solomon Shereshevskii, experience synesthesia: that is, the condition in which people "see" sounds or

"hear" colors. And while not all synesthetes seem to possess uncanny memories, people who have uncanny memories are often synesthetes. Mozart, for example, experienced a synesthesia in which he perceived music in colors.

41 **Walking is better than running:** Author Lyanda Lynn Haupt is dogged on this point. "Walking is the naturalist's pace—the tempo at which observation can occur," she writes on page 63 of *Rooted*. "We must be present, and we must be still. Then more still."

45 **The mind-work of memory:** I came across the term "mindwork" in reference to memory in environmental historian Lawrence Buell's essay, "Uses and Abuses of Environmental Memory," which appears in the anthology *Contesting Environmental Imaginaries*. It's perfect at communicating the fact that memory demands real effort.

47 **the calendar is a map:** Speaking of emergent ideas: I've been struck by how many writers in recent years have focused on the question of time as one of the fundamental handicaps of addressing the environmental polycrisis. It naturally comes up in Marcia Bjornerud's *Timefulness* and Jenny Odell's *Saving Time*. But it's also a major theme in Ben Ehrenreich's *Desert Notebooks*, where he writes (page 25): "The sky is a clock, and a calendar is also a map." He goes on to make a case for the necessity of exiting chronos's "messianic time." The French philosopher Bruno Latour once complained that modernity has "a peculiar propensity for understanding time that passes as if it were abolishing the past behind it."

48 **"the Great Acceleration":** In *The Great Acceleration*, authors John Robert McNeill and Peter Engelke describe this era as "an eccentric historical moment"—something never before experienced. They point out that less than one out of twelve people alive today can remember life before the acceleration began.

49 **a spiral is more like it:** To whom should I give credit? I feel like I plucked this out of the ether myself, working in the garden one day. But of course I didn't. It's an old idea, in Western philosophy at least as old as Hegel. Probably much older, for there are eight-hundred-year-old spiral pictographs chipped into red rock by the Pueblo cultures in the American Southwest. For a modern take, see Priya Subberwal, "The Shape of Time," *Orion*, March 18, 2025.

CHAPTER 2: BEAR WITNESS

52 **who had experienced the monarch diminishment as a steady decline:** Recent research confirms that people are more likely to perceive environmental change when it is presented as a binary rather as a gradual shift. In a large-scale cognitive experiment, subjects were presented with two sets of data: one showed temperatures rising steadily over time, and the other delivered the exact same information, only offered in the form of black-and-white data about whether a local lake froze or not. Perhaps not surprisingly, test subjects were better able to notice change when it was sharp. "The findings suggest that if scientists want to increase public urgency around climate change," a reporter at *Grist* wrote, "they should highlight clear, concrete shifts instead of slow moving targets." See: Kate Yoder, "Scientists Just Found a Way to Break Through Climate Apathy," *Grist*, May 5, 2025.

55 **aware of an awful irony at work:** Oliver Milman draws attention to this irony in his grim book *The Insect Crisis* when he writes (page 178), "It's a savagely remarkable dimension to the story of monarchs that the wonder they provoke appears to be doing little to shield them from being rubbed out. . . . Despite all the angst-ridden efforts, the monarch still faced being erased from all but a few pockets of North America."

57 **his memoir, *Speak, Memory*:** Nabokov's original title for the book was *Speak, Mnemosyne*. (Fortunately, someone talked him out of it.) In Greek myth, Mnemosyne was a Titan, the goddess of memory. She was the mother of the nine muses, who were the wellsprings of all art and culture. Among these was Clio, the muse of history. Clio's gifts of memory made her paramount among the muses. History came first, for without memory there's no poetry, or tragedy, or comedy, or song. Even at the very beginnings of Western history, the ancient Greeks understood that the past always beats within the present.

58 **"gaudy" in the words of one lepidopterist:** This characterization, along with the subsequent quip about the "crazy cat lady convention" came via Liam O'Brien, a former Broadway actor who invented a second career for himself as one of the San Francisco Bay Area's

most dogged butterfly conservationists. O'Brien's work to protect the Lange's metalmark butterfly is featured in Jon Mooallem's *Wild Ones*. It's yet another book in which shifting baseline syndrome has a cameo. Mooallem writes (page 129), "Acknowledging the problem of shifting baseline syndrome, like truly acknowledging the enormity of climate change, can be profoundly disturbing and discouraging. It begs the question of what baseline biologists should be measuring wildlife populations against in the first place."

60 **the spectacle inspires rhapsodies:** In her book about how citizen scientists and professional lepidopterists collaborated to decode the mysteries of the monarch migration, *Four Wings and a Prayer*, Sue Halpern writes (page 71), "The butterflies stream past like spawning salmon." Given the current state of salmon runs in much of North America, I think this line, now a quarter of a century old, is yet another example of shifting baseline syndrome. At one point Halpern describes the monarch as "the most common butterfly in North America." Today it would be more accurate to say "once common."

63 **modern environmentalism has become a technocratic argot:** I find the modern environmental movement's turn toward an officious reliance on law and science to be befuddling given environmentalism's roots in Romanticism and Transcendentalism. The canon of environmental literature—Thoreau, Muir, Leopold, Dillard—is packed with emotion. The rank-and-file environmentalists I have met in my lifetime often talk about their work in emotive terms. I think the environmental movement's reputation for being jammed with policy wonkery is a function of how NGOs—especially the "Big Green" groups—are so often forced to make their arguments in courts of law and official government proceedings. Over the years, this has led to an internalization of a cerebral, emotionally detached rhetoric.

64 **Feelings will prove one of the most effective:** Clayton Aldern makes virtually this exact same point in *The Weight of Nature*, when he writes (page 42), "Climate empathy, the antidote to climate amnesia, requires a dynamism that matches not just the structure of our own environments, but those of others and environments past as well."

65 **False memories are real things:** There's a disconcerting literature about how easily we create false memories, how we can go so far as

to make up entire episodes that never happened—or be coaxed by the suggestions of others into fabricating memories. Going down that wormhole is beyond the scope of my investigations. For more, see the work of researcher Elizabeth Loftus.

66 **Every generation believes:** As long ago as 380 BCE, Plato was complaining that the glory days were long gone. Already, the land had grown thin and wasted. In the *Critias* he describes primordial Athens in the millennia beyond memory: "In those days, the country was fair as now and yielded far more abundant produce. . . . Compared to what was, there remains only the bones of the wasted body . . . all the richer and softer parts of the soil having fallen away, and the mere skeleton of the land being left."

66 **a cautionary example of how emotion can twist memory:** In *How the Word Is Passed*, his book about Americans' efforts to remember (or forget by way of elisions) the history of slavery, Clint Smith visits Monticello, where a volunteer guide offers (page 41), "There's a difference between history and nostalgia, and somewhere between those two is memory." This strikes me as an insightful parsing. History alone is too cerebral; nostalgia is weighted down by emotion. In between those two poles one might find an honest sort of memory.

66 **Before nostalgia was a mood:** For a deep dive into nostalgia, see Svetlana Boym's encyclopedic *The Future of Nostalgia*. Among other insights—plus a sharply observed chapter on Nabokov's nostalgic tic—Boym makes the point in her Introduction that nostalgia is a reaction to modernity. "Somehow progress didn't cure nostalgia but exacerbated it."

67 **The fantasy is a mirage:** Ranganath sums it up nicely when he writes (page 48), "The cost of nostalgia is that it can leave us feeling disconnected from our lives in the present."

74 **where private grief is channeled into public mourning:** For more on the imperative of memory-making in an era of climate emergency, check out Lauren Markham's book, *Immemorial*. Markham notes (page 25) that "most memorials are made to remember the victims of human-made atrocities—war, genocide, mass shooters, terrorist attacks" and that "memorials are bound up with problems of power." Non-human beings are virtually powerless on a human-dominated planet, which is why ecological memorials are so few and small.

77 **Neuroscience has shown that stress, depression, and anxiety:** As Ranganath acknowledges in *Why We Remember* (page 92), "The effects of stress on memory are complicated." That said, it's well established that chronic insecurity and depression tax the memory systems. I recall a line that appears in Ehrenreich's *Desert Notebooks* (page 89), when he quotes a *New York Times* report saying that Trump's "tornado of news-making" had scrambled "Americans' grasp of time and memory."

77 **Joy, in contrast, energizes memory:** Proust, again, for the win. Recall the famous episode of the madeleine dipped in tea—it's an experience of joy. "Whence could it have come to me, this all-powerful joy?"

79 **McCarthy is insistent about the power of joy:** McCarthy, of course, isn't the only person to make such a claim. In recent years there's been a flourishing of writing about the political power of joy. A short list of American writers and activists who have made similar arguments would include Ross Gay, adrienne maree brown, Mary Annaïse Heglar, and Rebecca Solnit, among others. While I appreciate the celebration of joy as a form of political liberation, I think it's fair to say that a politics of joy also has its limits. Bouquets are no match for bulldozers.

CHAPTER 3: MAKE A RECORD

86 **Magnuson wrote in a paper published in 1990:** See, "Long-Term Ecological Research and the Invisible Present," *BioScience*, July–August 1990. I had a lovely conversation with Magnuson in December 2023, during which he expanded on his concept of the "invisible present" and how, in combination with "shifting baseline syndrome" and "the extinction of experience," it was a perfect example of an emergent idea. Magnuson told me that originally the term was a way of encapsulating the importance of long-term ecological research. "The idea of the invisible present was a way of having an elevator conversation with a politician or somebody making a decision about your grant." He was passionate about the importance of citizen science and ensuring that academic research is made easily available to the public—its own form of bearing witness. "It's not enough to collect data. You

have to make the data available, and you have to use that data to influence public decision making."

87 **The research done at National Science Foundation–funded LTER sites:** There are now a total of twenty-seven LTER locations, stretching from Arctic Alaska to Florida to Antarctica. The future of the LTER program, however, is in doubt due to Trump administration–proposed cuts to the National Science Foundation.

90 **I started keeping a nature-and-garden diary:** If beginning from a blank page seems too intimidating, I'd suggest Margaret Renkl's *Leaf, Cloud, Crow: A Weekly Backyard Journal*, which offers helpful prompts for what to record season by season.

91 **I have the receipts to prove it:** During my reporting in Wisconsin, I spent an afternoon with Gary and Penny Shackelford, a pair of retired doctors who are among the most impressive conservationists I have ever met. Over the decades, Gary and Penny have carefully and conscientiously—and at great expense—rehabilitated a former orchard south of Madison and restored it to prairie. Recently, they gifted the property, Fair Meadows, to the Southern Wisconsin Bird Alliance. They are also world-class phenologists; Gary has been keeping a detailed nature diary for several decades now. The entries are made digitally—so much for pen and pencil!—which has allowed him to index the entire corpus. Gary went so far as to teach himself a coding language to come up with a bespoke database. *Sandhill crane—go!* With AI swiftness, Gary can tell you the exact date, for every single year, on which the species has appeared at Fair Meadows on its spring migration.

97 **a whole academic discipline called historical ecology:** In the introduction to *Issues and Concepts in Historical Ecology*, the textbook editors write, "Historical ecology is neither a discipline nor a theory. It is a research framework." Point taken. But I think the distinction may be a bit too fine for non-academics.

101 **These kinds of historical records:** In the spring of 2025 I spent a couple of hours talking with Daniel Pauly on a bench on the University of British Columbia campus, the peaks of the Canadian Cascades poking out to the north, and he took some credit for the emergence of historical ecology. His original essay about shifting baseline syn-

drome had argued for "adding history" to a discipline—fisheries science—"that has suffered from lack of historical reflection." You can draw a straight line from those nineties-era complaints to today's robust field of historical ecology. There's an entire chapter on shifting baseline syndrome in *Issues and Concepts in Historical Ecology*. See also: James Dineen, "Looking More Deeply into the Past to Gauge the Planet's Future," *Undark*, May 18, 2020.

101 **"The big changes happened a long time ago":** I am reminded of a haunting observation from Barry Lopez, who wrote, "After the last undomesticated stretch of land is brought to heel, there will only records—strips of film and recording tape, computer printouts, magazine articles, books, laser-beam surveys—of these immensities. And then any tyrant can tell us what it meant, and in which direction we should go now. . . . And the naturalist . . . will be an orphan. He will become a dealer in myths." From "The Naturalist," *Orion*, October 1, 2001.

102 **Milan Kundera called "organized forgetting":** The phrase appears in Kundera's *The Book of Laughter and Forgetting*, and the full sentence is haunting: "Or is it true that the people will be unable to survive crossing the desert of organized forgetting?" The phrase has been widely borrowed, including by the speculative fiction writer Ursula K. Le Guin, who once used it in reference to the European notion of "discovery," which requires the fiction that no one knew the place before the colonizers showed up.

102 **put a gloss on the history of the January 6 insurrection:** The tawdry attempt to rewrite the events of January 6 are well known. Smithsonian Institution historians Jon Grinspan and Peter Manseau have made the case for how dangerous this is. "Nothing in our past, no matter how blatant it may seem to us today, is guaranteed eternal condemnation. . . . Enshrining rioters as heroes could be done fairly quickly." See, "How Will History Remember," *New York Times*, January 9, 2022.

102 **laws intended to constrict the teaching of American history:** Anti-fascist scholar Jason Stanley warns how dangerous this is in his book, *Erasing History,* when in the preface he writes: "By removing the uprisings from the curriculum, the authorities leave students

with the impression that the status quo has never been—and cannot be—challenged."

103 **nature is political:** In *Orwell's Roses*, Rebecca Solnit writes (page 29) that "Nature itself is immensely political. . . . The apoliticalness of nature [is] itself a political production." It's worth a detour to recall the political role that nature plays in Orwell's writings. Many of Orwell's fearful imaginings—Big Brother, doublethink, newsspeak—by now have become cultural shorthand. Less well remembered are the many ways in which nature is a vehicle for Winston's rebellion. Winston's first treasonous tryst with his lover occurs in the countryside, in a secluded grove "misty with bluebells" and full of dove song. His most prized possession is a piece of coral encased in glass, a paperweight. That fragment of nature is so precious because it resides outside the bounds of the police state. Again and again, Winston tries to return to his most beloved memory: the bucolic English countryside in the years before the Party subsumed everything. He thinks of this as the "Golden Country." The place, as Orwell describes it, is the apotheosis of a what today is considered now-lost pastoral England: "It was an old, rabbit-bitten pasture, with a foot track wandering across it and a molehill here and there."

105 **taking pictures of glaciers from the same viewpoint:** Paintings serve a similar function. In 2022, British landscape artist James Hart Dyke followed in the footsteps of the French artist Gabriel Loppé to recreate a painting of Switzerland's Mont Blanc. In the decades since Loppé's 1873 ascent, the scene had changed dramatically. "I couldn't but be aware of the huge changes that have happened across the years since," Dyke said. See: Joanna Moorhead, "Artist Captures the Impact of Climate Crisis over 150 Years on Mont Blanc," *The Guardian*, September 2, 2023.

106 **It's hard to argue with a picture:** Audio recordings can also provide indelible evidence of past environmental conditions. One of the most impressive repositories of natural soundscape audio recordings comes from Bernie Krause, who has spent decades capturing what he calls the "biophony"—the sounds of the living world. His vast recordings form an auditory account of how landscapes are being degraded; while natural sounds like birds and water have diminished,

artificial sounds like trains and airplanes have grown louder. I recommend Bernie's book, *Voices of the Wild* (Yale University Press; 2015). See also: Janelle Zara, "The Sound Ecologist Capturing a Disappearing World," *The Guardian*, June 14, 2023.

107 **a book titled *Exploring Glaciers—With a Camera*:** The small, instructive book by A. E. Harrison was published by Sierra Club Books in 1960, well before most people had heard of climate change. Even then, its author understood that glacier re-photography would have historical value: "The glacier story is always changing, but it can be caught and preserved with the proper use of a camera."

113 **sit unseen on data servers and hard drives:** Sorry to say that Andrew Fountain and Phil Rhodes are in good company. Many important photo archives are at risk of being lost. According to the *Washington Post*, "Without a clear plan, a photographer's archive can become a burden rather than a legacy, leaving heirs—spouses, children or executors—to navigate complex decisions about storage, access and preservation." See "Photos Are Disappearing, One Archive at a Time," *Washington Post*, March 10, 2025.

113 **If a glacier melts off a mountain:** Of course, Andrew Fountain's website isn't the only depository of glacier re-photography. The US Geological Survey also hosts several repeat photography sites to "quantify changes in glaciers over time." But most of us aren't going to be paid government scientists. The point here is that grassroots, community-driven natural history is more difficult than it may first appear.

CHAPTER 4: PASS IT ON

116 **Jewell had another name:** As Jewell told me, and as other Lummi fishermen echoed, the Lummi had it much easier than other Washington tribal nations during the height of the Fish Wars. The Nooksack River itself is incontrovertibly Lummi territory, and no one messed with them fishing the river. But whenever Lummi sought to fish in Bellingham Bay, at the river's mouth, they would face harassment.

117 **A Lummi collective that specializes in creating massive totem**

poles: Like the term "Indian," "totem pole" is an Anglo-European concoction, as Gregory P. Fields reports in *A Totem Pole History*, a book about the Lummi carver Joe Hillaire written by Pauline Hillaire. The term comes from the Anishinaabe word *ototeman,* which pertains to the idea of kinship. In that sense it's not entirely inappropriate.

119 **"telling us children's stories":** For a non-Native journalist, this cuts a bit close. As Lynne Kelly writes of Indigenous legends in *The Memory Code* (page 10), "The reason these stories appear simplistic is that they are usually the version told to children—and to cultural outsiders." There are, undoubtedly, many other Lummi tales of salmon that I will never know, being among the uninitiated.

121 **British navigator George Vancouver wrote:** In *The Good Rain*, Seattle-born Pulitzer Prize winner Timothy Egan notes (pages 88–89) that the Cascadian vistas seemed to have a liberating effect on Vancouver. He was a humorless man who was fond of flogging his sailors, "but when he entered Puget Sound something happened, as if he'd tossed his old spirit overboard in a rush of spring euphoria." He gave the crew a holiday and "his journals started to sing."

123 **How else to explain a celebratory:** See Lynda Mapes, "Big Salmon Run at Seattle's Ballard Locks Is Wowing Visitors," *Seattle Times*, August 18, 2023. I should say that Lynda is a friend, and I know that she understands the challenges of shifting baseline syndrome as well as anyone. Blame the headline writers.

124 **In *Salmon*, Mark Kurlansky spends:** Shifting baseline syndrome haunts the pages of Kurlansky's encyclopedic fish biography. "When you are cheering 4,000 fish in a river that used to produce well more than 100,000, this is what biologists call a 'shifting baseline,'" he writes on page 222, and then after the coda that I cite, dedicates the better part of his appendix to shifting baselines, which he explains as "the problem of the declining faith in nature."

125 **the reefnet being a unique fishing technology:** A reefnet works by creating an artificial reef that channels a school of salmon between two boats or floating platforms. As the fish pass the man-made reef, the crews on the platforms pull up a net and haul them in. It's considered especially sustainable because it has a high escapement percentage—that is, it allows many fish to continue onward to spawn.

The reefnet is the heart of Lummi culture. Reefnet sites were carefully guarded and passed along family lines. Jewell told me that the reefnet is "cosmologically female. . . . The mother gives you a solid foundation for life." In 2016, the late Lummi fisherman Larry Kinley resumed reefnetting for the first time in decades. "Even our reefnetting is coming full circle," Steve Solomon said with pride.

127 **federal judges have upheld Native treaty rights:** Federal courts, all the way up to the Supreme Court, have consistently ruled in favor of tribal treaty rights. A century ago, the Supreme Court, in a case over Yakama fishing rights, declared, "The treaty was not a grant of rights *to* the Indians, but a grant of rights *from* them." (My emphasis.) But all too often these court rulings went unenforced. In *To Fish in Common*, Daniel L. Boxberger offers an important insight when he writes (page 52), "Indians tend to win in the courts but lose in a practical sense."

129 ***"These are borrowed words"*:** This ethical idea is also expressed by members of the Western Apache, as Ken Basso relates (pages 29–30) in his ethnography, *Wisdom Sits in Places*: "Western Apache place-names were created by his ancestors . . . they were—and are—his ancestors' very own words. . . . Whenever one uses a place-name, even unthinkingly, one is quoting ancestral speech."

129 **I have a responsibility to pass it on:** I should step carefully here. I am not, it should be obvious, Native American. Cultural generalizations are dangerous, stereotypes are a trap, and as Indigenous scholars have pointed out, "Indian" is a fiction. You can no more describe the hundreds of distinct Native cultures of North America with a single word than you could capture the essence of both the Portuguese and the Russians with only the term "European." But I've listened carefully, and it seems to me that among the common threads linking diverse Indigenous cultures is a reverence for story.

130 **the most universal of human arts:** In *Why We Remember* (page 175) Ranganath reports: "According to one analysis, 40 percent of our conversational time is spent on storytelling and on creating and exchanging collective memories."

130 **clay tablets stamped with cuneiform are mere infants:** The ancients were suspicious of writing. In Egyptian myth, the god-king Theuth was unimpressed when the god Thamus realized he had

invented writing. The crib notes of script, he warned, would atrophy the muscle of memory. That tale comes from one of Plato's dialogues, *Phaedrus*, in which Socrates delivers a rant (which can be read as sarcastic) against the perils of the then-novel technology. "If men learn the art of writing," Socrates cautions, "it will implant forgetfulness in their souls."

130 **ecological baselines that stretch to time immemorial:** "Time immemorial" is a phrase that comes up frequently in conversations with Indigenous peoples. It's a bedeviling term, almost koan-like, for it refers to something from beyond the reach of memory or record that nevertheless persists.

133 **A story's meaning comes from its culture:** In an essay that is well known within the discipline of memory studies, a twentieth-century Italian historian, Alessandro Portelli, offers a defense of oral historians as an important resource for history and histography. "What is really important is that memory is not a passive depository of facts, but an active process of creation of meanings." Those meanings spring, inevitably, from the subjectivity of both speaker and listener. "Subjectivity is as much the business of history as are the more visible 'facts'." In this sense, subjectivity is a feature, not a bug, as it puts a spotlight on what is most important to the conversationalists. See Alessandro Portelli, "What Makes Oral History Different." See also Alistair Thomson, "Memory and Remembering in Oral History," which appears in *The Oxford Handbook of Oral History*.

133 **Many writers worry, with good reason:** As an author, I've often felt this myself, and I think it's a common anxiety among writers. Lauren Markham confesses to such doubts in *Immemorial* (page 98), "As a writer, I knew that words could sometimes feel like not only the antithesis of action, but also, worse, mechanisms of concealment for doing nothing at all." In *Earth's Wild Music*, environmental philosopher Kathleen Dean Moore confesses in her preface, "I will write my last nature essay on a planet that is less than half as song-graced and life-drenched as the one where I began to write. . . . [One] temptation is to stop writing altogether."

135 **half the work of oral history is listening:** One of the freakish char-

acteristics of modernity—in comparison with the preceding tens of thousands of years human experience—is the way in which the wisdom of elders has been devalued. As Daniel Pauly told me, "Until about one hundred and fifty years ago, it was common to go to the elders in your community for advice and guidance on all sorts of matters." No longer, of course. The respect for elders' wisdom could be considered another casualty of the Great Acceleration. With changes happening so quickly, it's perhaps easy to assume that earlier generations' life lessons are of limited value. I'm not sure that assumption is a smart one.

137 **we are hardwired for sharing stories:** It's significant that both Ranganath and Schacter end their neuroscience books by discussing the transmission of memories across generations via storytelling. Ranganath goes even further and makes the point (page 193) that elder-to-youth memory transmission "might not be unique to humans. In orcas (aka killer whales), another species in which lifespan extends well beyond the period of fertility, transmission of knowledge and traditions (hunting preferences, play behavior, even mate preferences) appears to be driven by postmenopausal females. . . . Orca grandmothers play a critical role in the survival of the pod." Indeed. The matriarch of the L Pod of the southern resident orcas of the Salish Sea—a female known as L28—is some forty years older than anyone else in the pod, perhaps one hundred years old. She routinely leads the pod all the way back to the San Francisco Bay's Golden Gate, presumably because she remembers the winter Chinook runs that once filled the rivers of the Sacramento Delta. Aldo Leopold famously urged people to "think like a mountain," but geological time is a mind-bender. Perhaps, though, we can think like an orca, and in doing so come to better appreciate the role of memory in species survival.

142 **passion for genealogy is universal:** In *Time and Water*, the Icelandic writer Andri Snær Magnason invites us to a thought experiment (based in his own family history) about how long memories can stretch. Let's say your grandmother was born in 1924; that she lives to the age of ninety-four; and that, sometime before she passes, she shares her earliest memories with your daughter (her great-

granddaughter), who was born in 2008. Now, imagine that your daughter also lives to ninety-four. And imagine that, before she dies, she hands down to her grandkids those stories she heard from her own great-grandmother when she was a child. If someone from that subsequent generation also lives to ninety-four, then the string of oral history would extend out to 2186. "You can touch 262 years with your bare hands," Magnason concludes. That's a period older the American Republic.

143 **"So the place isn't just, you know, *name of street*":** There's a world of difference—and a world of meaning—in the distinct ways that traditional cultures create place names. Many Indigenous toponyms are linked closely to the natural processes unique to that one site. So, for example, the Lummi name for today's village of West Sound on Orcas Island was *El e'ley*, "the place to cook clams." In *Wisdom Sits in Places*, Keith Basso writes (page 23), "What impresses me most of all is the rich descriptive imagery of Western Apache place names." *Tsee Ditl'ige Naaditine* means "trail extends across scorched rocks." Compare these Indigenous place-names with Western ones: Mount Baker, Puget Sound, Mount Rainier. Whereas Western place names are declarative—an act of appropriation via naming—Indigenous place names are descriptive. They seek to convey the essence of place. And even when Western place names are linked to the environment—how many thousands of creeks in North America are named "Bear Creek"?—they are more often linked to one single happening (a settler once saw a bear there) than to a recurrent process (the tide goes out and makes shellfish gathering easy). But go far back enough in time, and you'll discover that all sorts of preindustrial cultures knew, as the Western Apache still know, that wisdom sits in places. A famous example is Oxford, England—the oxen's ford, the place where the cattle cross the river.

146 **you have to go back to go forward:** On this point, see the venerable Standing Rock Sioux scholar Vine Deloria Jr., who in the preface to *The World We Used to Live In* writes, "We need to see where we have been before we see where we should go."

CONCLUSION: MEMORIES OF THE FUTURE

147 **bald eagles lived nearly everywhere:** The eagles made me do it—made me write this book, that is. Some years ago, my family and I were driving home to Bellingham after a vacation in the San Juan Islands when, crossing over what's called Colony Creek, a bald eagle flashed above the car. It was wintertime, the season when eagles congregate here in great numbers, and it wasn't an unusual sight. But it felt like the visitation of a muse: What would my descendants think of bald eagles, I wondered, some decades from now? Will they still be common, or perhaps threatened once again? How will the baselines shift over time? Those questions have preoccupied me ever since.

147 **The First Nations perceived:** My capsule history of the bald eagle story relies heavily on Jack Davis's encyclopedic wildlife biography, *The Bald Eagle*, which I highly recommend. I feel bad for not giving the author an explicit callout in the main text, but I worried it would distract from the legend vibe. I hope you understand, Jack!

148 **the unofficial national bird:** It wasn't until December 2024 that the bald eagle was officially designated as the United States' national bird.

148 **But it was a conflicted relationship:** As Davis details, by the turn of the last century, the emblem had eclipsed the eagle itself. One could find no shortage of places named for eagles, sports teams with eagle mascots, or newspapers with *Eagle* on the masthead—but the bald eagle itself was gone from the sky.

149 **Americans murdered eagles mercilessly:** Though it's hard to imagine today, the bald eagle was almost universally reviled. Even the Audubon Society considered the bird a pest that deserved to be shot. For a gripping recounting of the campaign by champions like Rachel Carson and Rosalie Edge to reform the Audubon Society's position on eagles, see Michelle Nijhuis's *Beloved Beasts*, Chapter 6, "The Eagle and the Whooping Crane."

150 **"little more than sewers to the sea":** The evocative line came from Senator Edmund Muskie. This point about the importance of the Clean Water Act to the successful revival of the bald eagle can't be overstated. When it comes to species recovery, habitat is everything.

If a wildlife species has no healthy home to return to, captive breeding and daydreams of de-extinction won't be enough to truly restore it.

154 **the gray whale return:** Mia said this to me in January 2025, and at that time it was very much still true. From the end of commercial whaling to the 2020s, the population of Eastern Pacific Gray Whales—the ones that migrate up and down the Pacific Coast of North America and which Mia knows best—enjoyed a steady increase. But in recent years the population has cratered. Marine biologists believe the decline is due to a diminishment of their food sources as ocean temperatures rise. In June 2025, NOAA Fisheries announced that the gray whale population was at its lowest level in fifty years. Even a wildlife comeback can suffer from downward shifting baselines.

155 **The indivisible present can cloak the march of progress:** I think this phenomenon of taking progress for granted is especially acute when it comes to social relations. Think about how quickly popular attitudes on gay marriage have shifted. In 1996, roughly 70 percent of Americans were opposed to gay marriage and only about 30 percent supported same-sex union; by 2025, the numbers had completely reversed, with some 68 percent of Americans expressing support for gay marriage. Think about American apartheid. My father grew up in the Jim Crow South, when many people considered segregation normal. For my daughter, segregation is nothing short of madness. Prejudice and racism are still alive and well, of course—but the progress we've experienced can be difficult to see.

155 **triumphs of harm avoided:** This is a common insight among environmentalists, who know all too well that every victory is provisional: the loggers only have to get into the forest once, but the trees have to be defended again and again. Solnit makes this point in *Hope in the Dark*, when she writes (page 74), "most environmental victories look like nothing happened. . . . They are triumphs invisible except through storytelling." *Hope in the Dark* also inspired some of my thinking of the role of history in progress. Solnit writes (page 34), "Often the road to the future leads through the past."

162 **Even fantasy is made:** I'll grant that every once in a while some genius alights on a truly original idea. But much of even the most outlandish fiction comes from fact. *Star Wars*? Famously, a space opera

set to ancient mythic archetypes, the fearsome AT-ATs inspired by the cranes at the Port of Oakland. *The Wizard of EarthSea*? A bildungsroman set at a boarding school. Even on an ever-changing planet, there is little that is new under the sun.

163 **John Lewis Gaddis writes:** If you're looking for a short, accessible book on histography, you can't do much better than Gaddis's book, which calls to mind the ideas about memory palaces. "What if we were to think of history as a kind of mapping? . . . The past is a landscape."

166 **Remembering is a form of resistance:** When Daniel Pauly and I met on the UBC campus in the spring of 2025, the conversation turned to the questions of the state of the planet and hope and despair. Pauly insisted to me one's emotional state or intellectual outlook is irrelevant. If you are optimistic about the future of the planet, then act to make it better still; if you are pessimistic about the future of the planet, then act to avoid the worst. Hope and despair are beside the point—no matter what, *act*.

167 **What's at stake in this work:** Many scholars of shifting baseline syndrome have made similar points, but perhaps none so insistently, and eloquently, as the psychologist Peter Kahn, who warns that shifting baseline syndrome could cause harm that lasts many human generations due to "unfulfilled flourishing." He asks us to "imagine a gray wolf held captive in a zoo, pacing from one end of its cage to the other repetitively for hours. It is hard to escape the interpretation that the wolf in a zoo is living an impoverished life." If we can't break free of environmental amnesia, we will be imprisoned and impoverished.

RECOMMENDED READINGS

Aguon, Julian. *No Country for Eight-Spot Butterflies.* New York: Astra, 2022.

Albrecht, Glenn A. *Earth Emotions.* Ithaca, NY: Cornell University Press, 2019.

Aldern, Clayton Page. *The Weight of Nature.* New York: Dutton, 2024.

Basso, Keith H. *Wisdom Sits in Places.* Albuquerque: University of New Mexico Press, 1996.

Benjamin, Walter. *Illuminations.* Edited by Hannah Arendt. Boston: Mariner, 2019.

Bjornerud, Marcia. *Timefulness.* Princeton, NJ: Princeton University Press, 2018.

Blackhawk, Ned. *The Rediscovery of America.* New Haven, CT: Yale University Press, 2023.

Blanchet, M. Wylie. *The Curve of Time.* Vancouver, BC: Whitecap Books, 2011.

Blumenthal, Richard W., editor. *With Vancouver in Island Washington Waters: Journals of 12 Crewmen, April–June 1792.* Jefferson, NC: McFarland & Company, 2007.

Bogard, Paul, editor. *Solastalgia.* Charlottesville, VA: University of Virginia Press, 2023.

Boxberger, Daniel L. *To Fish in Common.* Seattle: University of Washington Press, 2000.

Boym, Svetlana. *The Future of Nostalgia.* New York: Basic Books, 2001.

Brown, Bruce. *Mountain in the Clouds.* Seattle: University of Washington Press, 1995.

Carson, Rachel. *Silent Spring.* Boston: Mariner Books, 2002.

Carson, Rachel. *The Sense of Wonder.* New York: Harper Perennial, 1998.

Challenger, Melanie. *On Extinction.* Berkeley, CA: Counterpoint Press, 2021.

Children of the Setting Sun Productions. *Jesintel.* Seattle: University of Washington Press, 2023.

Cruikshank, Julie. *Do Glaciers Listen?* Vancouver: UBC Press, 2005.

Crumley, Carole L.; Lennartson, Tommy; Westin, Anna. *Issues and Concepts in Historical Ecology.* Cambridge: Cambridge University Press, 2018.

Davis, Jack E. *The Bald Eagle.* New York: Liveright, 2022.

Deloria, Vine Jr. *Indians of the Pacific Northwest.* Golden, CO: Fulcrum, 2012.

Deloria, Vine Jr. *The World We Used to Live In.* Wheat Ridge, CO: Fulcrum Publishing, 2006.

DeVoto, Bernard, editor. *The Journals of Lewis and Clark.* Boston: Mariner, 1997.

Diamond, Jared. *Collapse.* New York: Viking, 2005.

Dove, Michael R. *Hearsay Is Not Excluded.* New Haven, CT: Yale University Press, 2024.

Egan, Timothy. *The Good Rain.* New York: Vintage Books, 1990.

Ehrenreich, Ben. *Desert Notebooks.* Berkeley, CA: Counterpoint, 2020.

Emerson, Ralph Waldo. *The Major Prose.* Cambridge, MA: The Belknap Press, 2015.

Foster, Jonathan K. *Memory: A Very Short Introduction.* Oxford: Oxford University Press, 2009.

Fowles, John. *The Tree.* New York: Ecco, 2010.

Gaddis, John Lewis. *The Landscape of History.* Oxford: Oxford University Press, 2002.

Genova, Lisa. *Remember.* New York: Harmony Books, 2021.

Giroux, Henry A. *The Violence of Organized Forgetting.* San Francisco: City Lights Books, 2014.

Grenz, Jennifer. *Medicine Wheel for the Planet.* Minneapolis: University of Minnesota Press, 2024.

Halpern, Sue. *Four Wings and a Prayer.* New York: Vintage Books, 2001.

Hannibal, Mary Ellen. *Citizen Scientist.* New York: The Experiment, 2016.

Hanson, Thor. *Close to Home.* New York: Basic Books, 2025.

Haupt, Lyanda Lynn. *Rooted.* New York: Little, Brown Spark, 2021.

Hillaire, Pauline. *A Totem Pole History.* Lincoln: University of Nebraska Press, 2013.

Hoffer, Peter Charles. *Clio Among the Muses.* New York: New York University Press, 2014.

Kahn, Peter J. Jr.; Hasbach, Patricia H. *The Rediscovery of the Wild.* Boston: MIT Press, 2013.

Kelly, Lynne. *The Memory Code.* New York: Pegasus Books, 2017.

Kimmerer, Robin Wall. *Gathering Moss.* Corvallis: Oregon State University Press, 2003.

Kimmerer, Robin Wall. *The Serviceberry.* New York: Scribner, 2024.

Kingsolver, Barbara. *Flight Behavior.* New York: Harper Perennial, 2012.

Kittredge, William. *Taking Care.* Minneapolis, MN: Milkweed Editions, 1999.

Kurlansky, Mark. *Salmon.* Ventura, CA: Patagonia, 2020.

Leopold, Aldo. *A Sand County Almanac.* Oxford: Oxford University Press, 1987.

Lipsky, David. *The Parrot and the Igloo.* New York: Norton, 2023.

Lopez, Barry. *Embrace Fearlessly the Burning World.* New York: Random House, 2022.

Louv, Richard. *Last Child in the Woods.* Chapel Hill, NC: Algonquin, 2008.

McCarthy, Michael. *The Moth Snowstorm.* New York: New York Review of Books, 2015.

McGaugh, James L. *Memory and Emotion.* New York: Columbia University Press, 2003.

McGrath, Darryl. *Flight Paths.* Albany: State University of New York Press, 2016.

McNeill, J. R.; Engelke, Peter. *The Great Acceleration.* Cambridge, MA: Harvard University Press, 2014.

Macfarlane, Robert. *Is a River Alive?* New York: Norton, 2024.

Mackinnon, J. B. *The Once and Future World.* Boston: Houghton Mifflin 2013.

Magnason, Andri Snær. *On Time and Water.* Rochester, NY: Open Letter Books, 2021.

Mapes, Lynda. *The Trees Are Speaking.* Seattle: University of Washington Press, 2025.

Markham, Lauren. *Immemorial.* Berkeley, CA: Transit Books, 2025.

Miles, John C. *Impressions of the North Cascades.* Bellingham, WA: Chuckanut Editions, 2009.

Miller, Bruce Granville. *Be of Good Mind.* Vancouver: University of British Columbia Press, 2007.

Millet, Lydia. *We Loved It All.* New York: Norton, 2024.

Milman, Oliver. *The Insect Crisis.* New York: Norton, 2022.
Monbiot, George. *Feral.* London: Allen Lane, 2013.
Mooallem, Jon. *Wild Ones.* New York: Penguin, 2013.
Moore, Kathleen Dean. *Earth's Wild Music.* Berkeley, CA: Counterpoint, 2021.
Nabokov, Vladimir. *Speak, Memory.* New York: Vintage Books, 1989.
Nijhuis, Michelle. *Beloved Beasts.* New York: Norton, 2021.
Odell, Jenny. *How to Do Nothing.* Brooklyn, NY: Melville House, 2019.
Odell, Jenny. *Saving Time.* New York: Random House, 2023.
O'Kane, Trish. *Birding to Change the World.* New York: Ecco, 2024.
Orwell, George. *1984.* New York: New American Library, 1984 edition.
Ostrander, Madeleine. *At Home on an Unruly Planet.* New York: Henry Holt, 2022.
Pauly, Daniel. *Vanishing Fish.* Vancouver, BC: Greystone Books, 2019.
Proulx, Annie. *Fen, Bog & Swamp.* New York: Scribner, 2022.
Pyle, Robert Michael. *Nature Matrix.* Berkeley, CA: Counterpoint, 2020.
Pyle, Robert Michael. *The Thunder Tree.* New York: Houghton Mifflin, 1993.
Ranganath, Charan. *Why We Remember.* New York: Doubleday, 2024.
Read, Albert. *The Imagination Muscle.* New York: Union Square & Co., 2023.
Renkl, Margaret. *The Comfort of Crows.* New York: Spiegel & Grau, 2023.
Restak, Richard. *The Complete Guide to Memory.* New York: Skyhorse Publishing, 2022.
Ruben, David C. *Memory in Oral Traditions.* Oxford: Oxford University Press, 1995.
Sanders, Scott Russell. *The Way of Imagination.* Berkeley, CA: Counterpoint, 2020.
Sarris, Greg. *Becoming Story.* Berkeley, CA: Heyday, 2022.
Schacter, Daniel L. *Searching for Memory.* New York: Basic Books, 1996.
Schama, Simon. *Landscape and Memory.* New York: Vintage Books, 1996.
Smith, Clint. *How the Word Is Passed.* New York: Back Bay Books, 2021.
Solnit, Rebecca. *Hope in the Dark.* New York: Nation Books, 2004.
Solnit, Rebecca. *Orwell's Roses.* New York: Viking, 2021.
Stanley, Jason. *Erasing History.* New York: Atria, 2024.
Suttles, Wayne. *Coast Salish Essays.* Seattle: University of Washington Press, 1987.
Tan, Amy. *The Backyard Bird Chronicles.* New York: Alfred A. Knopf, 2024.

Thoreau, Henry David. *The Journal of Henry David Thoreau.* Salt Lake City, UT: Peregrine Smith Books, 1984.

Turcotte Edenholm, Stephanie. *The Perfect Place* (self-published).

Twohy, Patrick J. *Beginnings: A Meditation on Coast Salish Lifeways* (self-published).

Williams, Florence. *The Nature Fix.* New York: Norton, 2016.

Wood, Harriet Harvey, and Byatt, A. S., editors. *Memory: An Anthology.* London: Vintage Books, 2009.

Wray, Britt. *Generation Dread.* Toronto: Alfred A. Knopf Canada, 2022.

Van Campen, Cretien. *The Proust Effect.* Oxford: Oxford University Press, 2014.

Vizenor, Gerald, editor. *Survivance.* Lincoln: University of Nebraska Press, 2008.

Yates, Steve. *Orcas, Eagles & Kings.* Boca Raton, FL: Primavera Press, 1992.

Zeman, Adam. *The Shape of Things Unseen.* London: Bloomsbury, 2025.

ACKNOWLEDGMENTS

A book is a work of many hands.

This one never would have made it into its present form without the insight of my editor, Matt Weiland. Matt was able to see the shape of the book I needed to write before I could see it myself. He saw the necessity of focusing on prescriptions rather than dwelling on grim diagnosis. Without his wisdom, I doubt this book would've made it to print.

The work made it to Matt's desk thanks to my agents, Jonah Straus and Alec MacDonald. Alec came calling just as the idea for a book on shifting baseline syndrome began percolating in my head. After a string of "glowing rejections" that knocked me sideways, Jonah and Alec kept me going until we found a home for the book at Norton. I'm thankful for their faith.

I am grateful for the wonderful crew of other writers who read early drafts of the manuscript and provided me with invaluable feedback. Among other excellent notes, Cynthia Barnett helped me to see where the original draft was bogging down and how to speed it up. On walks and hikes, Ed Grumbine allowed me to think aloud and stress tested my ideas. Kathy Dean Moore encouraged me to rethink how I presented the frog in hot water metaphor and to reconsider some of my notions about biophilia. Ted O'Connell brought to the manuscript the sensibility of a novelist and songwriter—as well as the sharp pen of a college writing instructor. Clay Aldern's feedback honed my passages on neuroscience; his fingerprints are scattered across the pages. I'm indebted to all of them.

Many other friends helped sustain the work. My old pal Andrew Gaines provided notes on the introduction that helped make the writing friendlier to readers who aren't steeped in environmentalism. Cisco Sabin sent me text threads full of shifting baseline syndrome mind-bombs. Chris Chimenti made sure I stayed on deadlines. Paul Glowaski helped keep my spirits up. In ways big and small, fellow environmental writers like Adam Federman, Ben Goldfarb, Jonathan Hahn, Mark Hertsgaard, Krista Langlois, Geoff McGhee, Lynda Mapes, Lauren Markham, Emma Marris, Jeremy Miller, Aimee Nezhukumatathil, and Christopher Preston supplied writerly good vibes and good advice. Years ago—while editing a short essay of mine about shifting baseline syndrome for *High Country News* that became the kernel of this book—Michelle Nijhuis prodded me to focus on solutions, which spurred me to come up with my capsule antidote to environmental amnesia.

I am thankful to Jorie Graham for her kindness in letting me reprint the beginning of "Poem," whose first line has been an inspiration.

I'm indebted to my Lummi neighbors, who invited me into

their community and on whose land these words were written. The fantastic Freddie Lane (Sul-ka-dub) set up many of the conversations that made Chapter 4 possible; he is, truly, a boundary spanner. Darrell Hillaire and Raven Borsey read the first draft of the salmon chapter and provided positive feedback and important factual corrections. Hy'shqe!

Some generous hosts gave me space to sit and write, walk and think. I'm grateful to Kamala Tully and Noya Kansky at the Mesa Refuge, who let me sneak into an alum session at that wonderful place. During my time at the refuge I wrote the first draft of Chapter 2 in a lightning burst, cloistered in a shed overlooking the edge of the continent. Mary Swenson and Jesse Thurston were amazingly generous by lending me their cabin on Eliza Island, where the Salish Sea archipelago spread out before me as I wrote parts of Chapter 4.

The art of reporting a nonfiction book resembles an iceberg: most of the research remains below the waterline, unseen by the reader. I conducted many interviews with people who never made it into the main text of the manuscript, yet whose contributions were essential to the final body of work. I'm grateful to all of those who were generous with their time. They include: Bill Baccus, Hassan Basagic, Jack Bynum, Drew Cashman, Jessica Griffiths, Craig Hillman, Larry Horsfall, Isis Howard, Natalie Johnston, Jay Julius, Mike Larrabee, John Magnuson, Brenna Marsicek, Beth and Ken Martin, Ali Nelson, Liam O'Brien, Daniel Pauly, Emma Pelton, Jon Riedel, the late Howard Roth, Kurt Russo, Gary and Penny Shackelford, Sapna Sharma, Dave Sluys, Richard Solomon, Greg

Sass, Georgina Shafer, Jennifer Tice, and Rachel (Chedda) Wilson. I appreciate each of you.

In addition to Matt Weiland, I'm thankful for the pros at Norton who shepherded this book to publication, including Steve Colca, Julia Druskin, Rebecca Homiski, Erin Sinesky Lovett, Susan Sanfrey, and Sarahmay Wilkinson. Yumiko Gonzalez Rios helped me to navigate the ins and outs of the production process. Avery Hudson provided sharp copyedits on the manuscript and made some clutch saves.

Freelance fact checker and reporter Wudan Yan did an incredible job of fact-checking every sentence of the text and rescued me from countless embarrassments. Any errors that remain are mine alone.

Above all, I'm thankful to my family for their unconditional support. Robbie Greenberg and Lisa Rich served as cheerleaders who encouraged me and caterers who kept me well-fed. Mom mailed me regular clippings from the *Wall Street Journal* about anything adjacent to nature or environmental amnesia. Lily was a great sidekick on adventures near and far and did a good job feigning interest in her old man's more esoteric passions. Nell Greenberg has been a champion and counselor without equal. I'll always remember that it was Nell's patience, encouragement, understanding, and love that created the space for me to write these words. She made the whole thing possible.